Mineral Economics and Policy

This book provides an introduction to the field of mineral economics and its use in understanding the behavior of mineral commodity markets and in assessing both public and corporate policies in this important economic sector. The focus is on metal and non-metallic commodities rather than oil, coal, and other energy commodities.

The work draws on John Tilton's teaching experience over the last 30 years at the Colorado School of Mines and the Catholic University of Chile, as well as short courses for Rio Tinto and other mining companies. This is combined with the professional consulting and academic research of Juan Ignacio Guzmán over the past decade, in order to demonstrate the industry application of the economic principles described in the earlier chapters.

The book should be an ideal text for graduate and undergraduate students in the fields of mining engineering and natural resource economics and policy. It should also be of interest to professionals and investors in mining and commodity markets, and those undertaking continuing education in the mineral sector.

John E. Tilton is a Research Professor in the Division of Economics and Business and a University Professor Emeritus at the Colorado School of Mines, Golden, Colorado, USA. He is also a Professor in the Department of Mining Engineering at the Pontificia Universidad Católica de Chile as well as a Resources for the Future University Fellow.

Juan Ignacio Guzmán is Chief Executive Officer of GEM Ltda., an industrial engineering company that advises the mining industry in management and economics, and of BOAMINE SpA, provider of software solutions for the mining industry. He is also Assistant Professor in the Department of Mining Engineering, Pontificia Universidad Católica de Chile.

About Resources for the Future *and* RFF Press

Resources for the Future (RFF) improves environmental and natural resource policymaking worldwide through independent social science research of the highest caliber. Founded in 1952, RFF pioneered the application of economics as a tool for developing more effective policy about the use and conservation of natural resources. Its scholars continue to employ social science methods to analyze critical issues concerning pollution control, energy policy, land and water use, hazardous waste, climate change, biodiversity, and the environmental challenges of developing countries.

RFF Press supports the mission of RFF by publishing book-length works that present a broad range of approaches to the study of natural resources and the environment. Its authors and editors include RFF staff, researchers from the larger academic and policy communities, and journalists. Audiences for publications by RFF Press include all of the participants in the policymaking process—scholars, the media, advocacy groups, NGOs, professionals in business and government, and the public.

Mineral Economics and Policy

John E. Tilton and Juan Ignacio Guzmán

First published 2016
by RFF Press
Routledge, 711 Third Avenue, New York, NY 10017

and by RFF Press
Routledge, 2 Park Square, Milton Park, Abingdon, Oxon OX14 4RN

RFF Press is an imprint of the Taylor & Francis Group, an informa business

British Library Cataloguing-in-Publication Data
A catalogue record for this book is available from the British Library

Library of Congress Cataloging in Publication Data
Names: Tilton, John E., author. | Guzmán, Juan Ignacio, author.Title:
Mineral economics and policy / John E. Tilton and Juan Ignacio Guzmán.
Description: Milton Park, Abingdon, Oxon ; New York, NY : RFF Press,
2016. | Includes bibliographical references and index.Identifiers: LCCN
2015044014| ISBN 9781617260889 (hbk) | ISBN 9781138838956 (pbk) |
ISBN 9781315733708 (ebk)Subjects: LCSH: Mineral industries. | Mines
and mineral resources--Economic aspects. | Mines and mineral resources--
Government policy.Classification: LCC HD9506.A2 T523 2016 | DDC
333.8/5--dc23LC record available at http://lccn.loc.gov/2015044014

ISBN: 978-1-61726-088-9 (hbk)
ISBN: 978-1-138-83895-6 (pbk)
ISBN: 978-1-315-73370-8 (ebk)

Typeset in Times New Roman
by Saxon Graphics Ltd, Derby

Contents

Figures

Tables

Preface

The origins of this book date back to the fall of 1972, when I joined the Department of Mineral Economics at the Pennsylvania State University. A decade or so earlier I had written my PhD dissertation on the determinants of international trade patterns for a number of major nonferrous metals. But only at Penn State did I start teaching a course on metal markets and industries, and so thinking about the metal industries and the field of mineral economics in a comprehensive manner. How do the supply and demand for mineral commodities differ between the short run and the long run? And, how do they differ from the supply and demand for agricultural or manufactured products? What can one say about the behavior of mineral prices and markets—why are they so volatile, for example? What important issues for public policy arise from the mineral sector, and what insights, if any, can simple economic tools shed on these issues?

Back then, governments worried about strategic and critical materials, particularly those such as cobalt and manganese largely coming from southern Africa and the Soviet Union. Depletion and the long-run availability of mineral commodities was another widespread concern, particularly in light of their non-renewable nature and the expected growth in the global economy. Even before OPEC (the Organization of Petroleum Exporting Countries) pushed up oil prices in 1973, cartels, monopolies, and mergers and acquisitions attracted considerable attention. While sustainable development was not an issue, certainly the pollution and environmental damage associated with mining and mineral production were.

Since 1972 I have continued to teach my course on mineral economics with its focus on metal and other non-energy mineral commodities. Over time the course has evolved as my teaching shifted from Penn State to the Colorado School of Mines and more recently to the Catholic University of Chile. The changes reflect in part the waxing and waning of the policy issues clamoring for public attention associated with mining and the mineral sector. They also reflect my own understanding of the mineral industries and the important economic, institutional, and technological forces governing their behavior.

Few academics have the opportunity to develop and teach over several decades their own specialized course. To my knowledge my course is unique with no close substitutes. For this reason, a decade or so ago, about the time I officially retired from the Colorado School of Mines, I began thinking about writing this book.

When I really retired, the book would provide those with an interest in mineral economics access to the conceptual framework and other material that I have found useful for understanding mineral commodity markets.

As the years slipped by with only a few chapters emerging into daylight, I began to realize I needed help from a good disciplinarian. It was then that I enlisted Juan Ignacio Guzmán to join me in this endeavor. I met Juan Ignacio while serving on his PhD dissertation committee at the Catholic University of Chile in 2007 and was impressed with both his command over the modern tools of economics and his in-depth understanding of copper and other commodity markets. Upon completing his dissertation, Juan Ignacio worked for several years in the Santiago office of CRU, a well-known consulting group focusing on mineral commodity markets. He then started his own, very successful consulting company, GEM (Gestión y Economía Ltda.). While working full-time, he has also continued to teach part-time at several universities in Santiago, including as a colleague at the Catholic University of Chile.

Juan Ignacio is the ideal partner. He complements my background with his strong knowledge of recent theoretical and quantitative developments in the field of economics, his interest in resource economics and industrial organization, and his understanding of mineral commodity markets based on extensive experience as an industry consultant. While my interests largely focus on appropriate public or government policies for the mineral sector, Juan Ignacio brings a strong interest in private or firm policies as well. Finally, let me also confess that, should this book find a useful niche helping interested readers better understand the fascinating machinations of metal and mineral commodity markets, I am hoping Juan Ignacio will from time to time provide updated and revised editions of this volume, as the importance of the mineral sector for welfare of humanity is not likely to wane over time.

Let me finish this preface by acknowledging and thanking some of those who helped make this volume possible. A complete list here is impossible, since we both have benefited over the years from the knowledge and insights of the many students we have taught and advised, of our colleagues, and of the engineers and other professionals with whom we have had the opportunity to interact. There are a few names, however, that we must mention.

First, we want to thank Tim Hardwick, our Senior Editor at RFF Press, for his enthusiasm and gentle prodding over the lengthy gestation period of this project, and Ashley Wright, Hannah Champney, Dave Wright, and Gary Smith for their assistance in the subsequent editorial and production processes. In addition, we are grateful to Robert D. Cairns, Phillip Crowson, Graham A. Davis, Roderick G. Eggert, Joaquín Jara, David Humphreys, Gustavo Lagos, Philip Maxwell, and Marian Radetzki for their useful insights and helpful comments on earlier drafts.

John E. Tilton
Golden, Colorado
September 30, 2015

1 Introduction

The men and women who extract mineral resources from the earth's crust and process them into a variety of different metals and other mineral commodities make life as we know it possible. It has not always been so. For millennia our distant ancestors lived off the land, the surface of the land, first from hunting and gathering, and then from raising their own crops and animals. Only about 40,000 years ago did this change when humans began on a significant scale extracting flint and useful minerals from within the earth. The subsequent division of history into the Stone Age, Bronze Age, and Iron Age highlights the importance of mining from that time onward.

Moreover, over the past several centuries the impact of mineral commodities on society has exploded. This period has witnessed the discovery and commercialization of many new materials, including aluminum, steel, chromium, lithium, and the rare earth minerals. Even taking into account the materials widely used as far back as the Roman era—copper, tin, lead, stones, and clays, for example—it is hard to think of a single mineral commodity whose consumption over the past 50 years does not exceed by a wide margin the quantities used throughout all previous history. These resources from the earth are needed to build homes, roads, bridges, offices, schools, and hospitals. They are found in computers, food and beverage containers, automobiles, airplanes, modern windmills, and the lowly 50-watt light bulb. Without them, modern industry, agriculture, communications, transportation, science, and medicine would simply not exist.

Mineral Economics and Policy

This book—*Mineral Economics and Policy*—provides an overview of how mineral markets and industries operate and why they behave as they do. The two of us together have devoted more than half a century to teaching, researching, and consulting on economic and policy issues associated with mining and with the production and use of mineral commodities. *Mineral Economics and Policy* draws on this work, synthesizing, updating, distilling, and extending it. It is written for both the specialist and the interested layperson and requires only an introductory understanding of supply, demand, and other basic economic principles. While far from comprehensive in the sense of treating all interesting economic and policy issues associated with the mineral industries and markets, the book does provide a conceptual framework that many of our students and we have found helpful. It then applies this framework to examine some of the interesting policy issues arising from the mining, processing, and use of mineral commodities.

The focus is on the metals and to a somewhat lesser extent the non-metallics. Among the metals some will detect a bias in favor of copper. This simply reflects our own interests over the years and the fact that both of us teach in the Mining Engineering Department at the Catholic University of Chile. Copper is tremendously important in Chile, accounting in recent years for nearly half of the country's export earnings and a large share of its government revenues as well. Nevertheless, much of the analysis applies as well to mineral commodities more generally, including diamonds, phosphate rock, and even petroleum. In this respect, *Mineral Economics and Policy* provides an introduction to the field of mineral economics.

Defined broadly, as is usually the case in North America, mineral economics covers three classes of commodities—energy, non-metallics, and metals. The energy minerals include coal, petroleum, natural gas, and uranium, and are by far the most important of the three groups in terms of sales. They are also the most studied. Many good economic analyses of energy markets exist, including several textbooks on energy economics.[1] By comparison, the metals and the non-metallic mineral commodities have received far less attention.[2]

Among the metals, iron and steel, aluminum, copper, gold, nickel, lead, zinc, and tin are the most important, at least in terms of sales (see Table 1.1). Many others also exist, whose importance should not be underestimated. Modern electronics, for example, is hard to imagine without semiconductors made of silicon or without liquid crystal displays and other flat-panel devices made with indium.

The non-metallics include construction materials (such as sand and gravel, stone and marble), fertilizers (such as potash and phosphate rock), and all the other mineral commodities that are neither metals nor energy resources (see Table 1.1). Their production when measured by volume or tonnage exceeds that of the metals. Nevertheless, the non-metallics have received even less attention from policy analysts and others than the metals. Just why this is so is not entirely clear. Part of the explanation may lie in the fact that they are for the most part large-volume, low-cost products often sold in regional or local markets. They enter international trade on a much smaller scale, and typically require less sophisticated technologies to extract and process. This, though, is far from a satisfactory explanation, and in any case numerous non-metallic products do not fit this simple generalization, as phosphate rock and diamonds so amply illustrate.

The mineral industries, it is important to highlight, produce a vast array of quite different substances and products. Some mineral commodities are extracted from large, open pits; others are hoisted out of underground mines; still others are pumped from deep wells; and a few are processed from seawater. Production can be simple and cheap, though often it is technically complicated and costly. Some mineral commodities are recovered as by-products or co-products in the course of extracting other commodities. Some are obtained from recycling old and new scrap. Some are found in only a few locations and are traded worldwide. Others are produced in many locations and consumed close to their terrestrial source. Some are traded on competitive exchanges where prices fluctuate daily. Others are produced by only a few firms and are offered at stable producer prices.

Mineral Economics and Policy does examine non-metallic mineral commodities, though it too is to some extent guilty of paying insufficient attention to the non-metallics, perpetuating an injustice to an important group of mineral commodities. The reason is simply that our own interests, as noted, have over the years largely focused on the metals. Fortunately, much of the conceptual framework and many of the policy issues explored in *Mineral Economics and Policy* are just as relevant and useful for the non-metallics as the metals.

This rich diversity makes mineral commodities and the field of mineral economics fascinating to study, but it also means no general model or economic analysis is applicable to all mineral commodities or even all the metals. Rather, each must be considered individually, thereby allowing the analysis to take account of its particular features.

As a result, a single book cannot begin to cover all the metals, let alone all mineral commodities, and no attempt is made to do so here. Instead, *Mineral Economics and Policy* concentrates on illustrating the usefulness of relatively simple economic principles, particularly those associated with supply and demand, in understanding the behavior of commodity markets.

A good mineral economist must possess a strong understanding of the technological and institutional conditions that shape and constrain the behavior of mineral commodity markets. What are the stages in the production of steel? How does mining law vary around the world and how, in turn, does it affect exploration? How capital intensive is the production of aluminum? How are the creation and diffusion of new technologies, such as the solvent extraction electrowinning process, altering the production and price of copper and other mineral products? How are they changing the associated environmental pollution?

For this reason, we believe mineral economics is more than just a specialty within economics. It is an interdisciplinary field that draws as well on geology, mining engineering, environmental science, metallurgy, and other associated technical fields.

There is one particularly important conclusion that the pages that follow highlight time and time again. The simple tools of economics can provide powerful insights into the nature and behavior of the mineral markets, but only if the analyst applying these tools has a firm understanding of the important technological and institutional relationships governing the market he or she is examining, and so can tailor the analysis to take these relationships explicitly into account.

The Road Ahead

Mineral Economics and Policy is divided into two parts. Part I develops the conceptual framework. Specifically, it assesses mineral commodity demand, supply, markets, and prices. In the process, it explores the nature of material substitution, recycling, and secondary production, along with by-product and co-product output. It analyzes the important forces shaping demand in both the past and the future. It examines the relationships between production costs, technological change, and supply. It describes how mineral commodity markets have evolved over the past half century and investigates the volatility of their prices in the short run and their secular trends over the long run.

Part II explores policy issues associated with mining, mineral processing, and use that are of particular interest to governments and other entities concerned with the welfare of society in general. In particular, it considers:

• How mining and metal production produces economic rents and the role of taxation and other policies in the distribution of this wealth.
• The reasons why certain countries enjoy a comparative advantage or competitiveness in the production and export of iron ore or other mineral commodities, while other countries rely on imports for all or most of their needs.
• The nature of market power and the evolution of antitrust or competitiveness policies in the mineral sector over time.
• The troubling relationship between mineral wealth and economic development, including the nature of the Dutch Disease and the arguments both for and against the Resource Curse.
• The long-run threat from mineral depletion and the important differences between the major mental models used to assess this threat—the fixed-stock paradigm and the opportunity-cost paradigm.
• Environmental and sustainability issues—such as intergenerational equity and the socially optimal amount of recycling, conservation, and substitution of renewable for non-renewable resources—associated with mineral production and use.

Notes

1 For example, see Dahl (2014).
2 There are fortunately a few good exceptions. See Crowson (2008), Humphreys (2015), Maxwell (2013), and Radetzki (2008). Though these, we believe, are good complements rather than substitutes for this volume.

References

Crowson, P., 2008. *Mining Unearthed*, Aspermont, London.
Dahl, C.A., 2014. *International Energy Markets*, revised edition, Pennwell Press, Tulsa, OK.
Humphreys, D., 2015. *The Remaking of the Mining Industry,* Palgrave Macmillan, Houndmills, Hampshire, UK.
Maxwell, P., 2013. *Mineral Economics*, second edition, Australasian Institute of Mining & Metallurgy, Carlton, Victoria.
Radetzki, M., 2008. *A Handbook of Primary Commodities in the Global Economy*, Cambridge University Press, Cambridge.

Table 1.1 Classification, end products, and 2014 sales in millions of US dollars for selected mineral commodities

Classification		Commodity	Products	2014 sales [million $]
Metals	Ferrous metals*	Iron ore	Concentrates Coarse ores Fines ores Pellets Briquettes Sinter Roasted iron pyrites	325,220
	Nonferrous metals	Base	Aluminum	Unwrought (in coils) Unwrought (other than aluminum alloys) Unwrought (billet) Bauxite, calcined Bauxite, crude dry (metallurgical grade) Alumina Aluminum hydroxide Waste and scrap	120,238
			Copper	Copper ores and concentrates Unrefined copper anode Refined and alloys (unwrought) Copper wire (rod)	129,862
			Tin	Unwrought tin, not alloyed Unwrought tin, containing lead Tin waste and scrap .	6,395
			Nickel	Nickel oxides, chemical grade Ferronickel Unwrought nickel, not alloyed	40,471

Table 1.1 *continued*

Classification		Commodity	Products	2014 sales [million $]
Metals	Nonferrous metals			
	Base	Lead	Unwrought (refined) Antimonial lead Alloys of lead	10,352
		Zinc	Zinc ores and concentrates Zinc oxide and zinc peroxide Unwrought zinc, not alloyed Zinc alloys	28,881
	Light	Berylium	Berylium ores and concentrates Berylium oxide and hydroxide Berylium-copper master alloy Berylium unwrought, including powders Waste and scrap	121
		Lithium	Lithium oxide and hydroxide Lithium carbonate Concentrates and others	1,265
		Titanium	Titanium oxides (unfinished TiO$_2$ pigments) TiO$_2$ pigments (80 percent or more TiO$_2$) Ferrotitanium and ferrosilicon titanium Unwrought titanium metal Other titanium metal articles Wrought titanium metal Synthetic rutile Ilmenite and ilmenite sand Rutile concentrate Titanium slag Titanium waste and scrap	4,982

Classification		Commodity	Products	2014 sales [million $]
Metals	Nonferrous metals			
		Platinum group metal (PGM)	Iridium	Iridium unwrought, including powders	582
			Palladium	Unwrought and semi-manufactured	5,070
		Precious	Gold	Unwrought gold, including bullion and doré	116,778
			Platinum	Unwrought and semi-manufactured	7,454
			Silver	Unrefined silver / Refined bullion	15,969
		Refractory	Cobalt	Cobalt ores and concentrates	3,457
				Chemical compound:	
				Cobalt oxides and hydroxides	
				Cobalt chlorides	
				Cobalt sulfates	
				Cobalt carbonates	
				Cobalt acetates	
				Unwrought cobalt alloys	
				Cobalt mattes and other intermediate products, cobalt powders	
				Wrought cobalt and cobalt articles	
				Waste and scrap	

Table 1.1 continued

Classification			Commodity	Products	2014 sales [million $]

Metals	Nonferrous metals	Refractory	Molybdenum	Molybdenum ore and concentrates
				Molybdenum chemicals
				  Molybdenum oxides and hydroxides
				  Molybdates of ammonium
				  Molybdates, all alloys
				Molybdenum pigments
				Ferromolybdenum
				Molybdenum metals
				  Powder
				  Unwrought
				  Wrought planes, sheets, trips, etc.
				  Wire
				Waste and scrap
					7,155

			Niobium	Synthetic tantalium–niobium concentrates Niobium ores and concentrates Niobium oxides Ferroniobium Niobium unwrought, alloys, metal, powders Waste and scrap	3,812
			Rhenium	Salts of peroxometallic acids Rhenium unwrought, powders Rhenium wrought Waste and scrap	146
			Tantalum	Synthetic tantalum–niobium concentrates Tantalum ores and concentrates Tantalum oxides Unwrought: powders and metal alloys Waste and scrap	355

Classification		Commodity	Products	2014 sales [million $]
Metals	Nonferrous metals	Refractory	Tungsten	Tungsten ores and concentrates Tungsten oxides Ammonium tungstates Tungsten carbides Ferrotungsten Tungsten powders	3,637
			Zirconium	Zirconium ores and concentrates Germanium oxides and zirconium dioxide Ferrozirconium Unwrought and zirconium powders Other zirconium articles Waste and scrap	1,494
		Nonclassified	Arsenic	Metal Acid Trioxide Sulfide	30
			Bismuth	Bismuth and articles thereof Waste and scrap	213
			Cadmium	Cadmium oxide Cadmium sulfide Pigments and preparations based on cadmium compounds Unwrought cadmium and powders Wrought cadmium and other articles Waste and scrap	43
			Chromium	Ore and concentrates Ferrochromium Ferrochromium silicon Chromium metal, unwrought, powder Waste and scrap	6,467

Table 1.1 *continued*

Classification		Commodity	Products	2014 sales [million $]
Metals	Nonferrous metals	Nonclassified	Indium	Unwrought indium, including powders	582
		Manganese	Ores and concentrates	36
			Manganese dioxide	
			High-carbon ferromanganeses	
			Silicomanganese	
			Metal, unwrought	
		Mercury	Mercury	100
		Silicon	Silicon	26,530
			Ferrosilicon	
			Ferrosilicon manganese	
		Strontium	Celestite	16
			Strontium metal	
			Strontium oxide, hydroxide, peroxide	
			Strontium nitrate	
			Strontium carbonate	
		Vanadium	Vanadium pentoxide anhydride	1,780
			Vanadium oxides and hydroxides	
			Vanadates	
			Ferrovanadium	
			Vanadium and articles thereof	
Non-metals	Construction		Cement	Cement clinker	411,730
			White portland cement	
			Aluminous cement	
			Other hydraulic cement	

Classification	Commodity	Products	2014 sales [million $]
Non-metals	Construction	Gypsum, anhydrite	2,214
	Salt	Salt (sodium chloride)	21,957
	Sulfur	Crude or unrefined Sublimed or precipitated Sulfur acid	6,878
	Fertilizers	Natural calcium phosphates	19,800
	Phosphate rock		
	Potash	Potassium nitrate Potassium chloride Potassium sulfate Potassic fertilizers Potassium–sodium nitrate mixtures	25,550
	Refractory	Crystalline flakes (not including flakes dust) Powders	1,480
	Graphite (natural)		
	Abrasive	Industrial miners' diamonds, carbonados Industrial diamonds, simply sawn, cleaved, or bruted Industrial diamonds, not worked Grit or dust and powder of natural or synthetic diamonds	800
	Industrial diamond		
	Insulate	Crocidolite Amosite Chrysotile crudes or milled fibers	3,089
	Asbestos		

Table 1.1 *continued*

Classification	Commodity		Products	2014 sales [million $]
Non-metals	Barite	Pigment	Ground barite Crude barite Oxide, hydroxide, and peroxide Chlorides or sulfates or barium Carbonate	1,158
	Diamond	Gemstones	Diamond, unworked or sawn	25,000
		Precious		
	Germanium	Nonclassified	Germanium oxides Metal, unwrought Metal, powder Metal, wrought	314
	Helium		Helium	688

Source: GEM, based on information from the US Geological Survey.

Note:

*There is no uniform classification for the ferrous metals. Here we define the ferrous metals as those containing iron. They include iron, pig iron, cast iron, wrought iron, and steel alloys. Other classification systems consider manganese, cobalt, chromium, and other metals used in producing steel alloys as ferrous metals.

Part I

Mineral Economics

A Conceptual Framework

2 Demand

The nature of demand varies greatly from one mineral commodity to another. For example, lead is largely consumed in the production of automobile batteries, while molybdenum is used primarily as an alloy to enhance the hardness, strength, toughness, wear, and corrosion resistance of steels and superalloys. Iodine is a common disinfectant and has numerous other medical and chemical applications, as it is a good absorber of X-rays, has low toxicity, and attaches readily to organic compounds. While such differences are interesting and can be important, an attempt to explore them in any comprehensive manner would soon leave us hopelessly submerged in a sea of endless details. This chapter focuses instead on the concepts and characteristics of demand common to most or all mineral commodities.

The chapter begins by considering measures of demand, and then turns to the major determinants of demand. Next, it examines the demand function, the demand curve, and demand elasticities—all frequently used economic concepts that can provide useful insights into the nature of mineral commodity demand. This leads to a discussion of the intensity-of-use and material substitution. The chapter then explores some interesting historical changes in commodity demand and concludes by highlighting the perceived wisdom in the field of mineral economics regarding the nature of demand for metals and other mineral commodities.

Measuring Demand

There are many different types of demand. We can, for example, distinguish among the demand for aluminum from an individual company (the Coca-Cola Bottling Company Consolidated of Charlotte, North Carolina), from an industry (the beverage container industry), or from a particular region or country (Latin America or Japan). On the sellers' side of the market, there is the demand for aluminum for a particular producing company (Alcoa) or a particular producing country (United States). Alternatively, one can identify the demand for aluminum from fabricators and other metal consumers for inventories, from commodity funds for investment and speculation, or from governments or international organizations for strategic and economic stockpiles.

Mineral economists, however, separate the various types of demand into two groups—market demand and all the rest. Market demand is the quantity that all the participants—fabricators, speculators, government agencies, and so on—in the relevant market are prepared to buy given the market price and other prevailing conditions. We distinguish market demand from all other types of demand because it is market demand that ultimately, along with market supply, determines market price and output (see Chapter 4). Other types of demand are of interest largely because they tell us something about market demand, and thus allow us to say something useful about market prices and production. For example, an analysis that separately considers the demand for lead in batteries for new automobiles, in replacement batteries for existing automobiles, and in other end uses may provide some interesting insights into the likely future trends for total lead demand.

Defining the Relevant Market

The relevant market encompasses all buyers and sellers whose interactions determine the market price and output. It possesses both a geographic and a product dimension. Many commodity markets, including those for iodine and gold, are global. The actions of buyers and sellers in Hong Kong, São Paulo, Amsterdam, New York, Dubai, and elsewhere around the globe determine either a single world price or a set of prices that move closely in step with each other. There are, though, other mineral commodities, such as crushed stone and salt, whose markets are regional or even local. Within a single country, such as the United States, many separate markets for crushed stone exist with quite different prices at any given time.

A market should similarly only include those participants that buy and sell the same product or very similar products. In the case of iron ore, for example, we often assume that pellets, high-grade lump ore, and sinter are close substitutes, and so those who trade these products belong together in a single iron-ore market.[1] Yet, this need not always be the case; depending on the purpose of the inquiry it may be more appropriate to define separate markets for pellets, lump ore, and sinter. For this and other reasons—see, for example, Box 2.1—just how the relevant market should be defined is not always clear.

Box 2.1 The "Insulated Copper Wire and Cable Scrap Suitable for Chopping" Market

A number of years ago, one of us (John Tilton) was involved as an expert witness in a civil antitrust case in the United States. Diversified Industries, a relatively small firm recycling copper wire, was suing AT&T. At the time, AT&T was a government-regulated monopoly and provided almost all of the telephone service in the country. As a result, it also controlled almost all scrap communication wire available for recycling. For years it sold this scrap to Diversified Industries and other companies. They chopped the insulated wire into small pieces, which freed the copper from its plastic or paper insulation, and then employed a gravity separation process to remove the valuable copper from the worthless insulation.

When AT&T decided to construct its own recycling facilities, Diversified Industries accused the company of monopolizing the "insulated copper wire and cable scrap suitable for chopping" industry and, as US antitrust law permits, sued AT&T for triple the amount of the damages it claimed to have suffered. In court, AT&T responded that the price of refined copper metal, which is set on Comex (the New York Commodity Exchange) and the LME (the London Metal Exchange), and which no one controls, determines the price of copper scrap, including the price of insulated copper wire and cable scrap. Since AT&T could not artificially raise the price of insulated copper wire and cable scrap by withholding supply, it did not possess any market or monopoly power. Diversified Industries did not challenge this argument, but responded by noting that a scrap processor could not use its chopping machines to recycle refined copper or even other types of copper scrap, and as a result these copper products were not a substitute for insulated copper wire and cable scrap.

The jury, after a long court battle, agreed with Diversified Industries, which as a result received many millions of dollars from AT&T. The case illustrates how difficult it can be to define the appropriate metal market. For most purposes, one is not likely to identify a separate "insulated copper wire and cable scrap suitable for chopping" market.

Conceptual Issues

After determining the relevant market, one still must decide the following.

1 At Which Stage of Production to Measure Demand

As Figure 2.1 illustrates, in the case of copper, we can measure: (1) the demand of smelters and refiners for copper in concentrates; (2) the demand of smelters and refiners for copper in scrap; (3) the demand of fabricators for copper in the form of refined copper; (4) the demand of fabricators for copper in the scrap they use directly in making ingot and mill products; (5) the demand of fabricators for copper in both refined metal and scrap; (6) the demand of manufacturers for copper in fabricated and semi-fabricated products; and (7) the demand of consumers for copper in the final goods they purchase. In practice, data may not be available to measure demand at all stages of production. Often, for example, we would like to know the demand of consumers for the copper and other materials in the final products they purchase and use, but this information is rarely available.[2] So we use instead the demand of fabricators for copper in refined metal and scrap, which normally is available. When doing so, however, we need to be aware we are using a proxy that may or may not closely approximate the data we actually want.

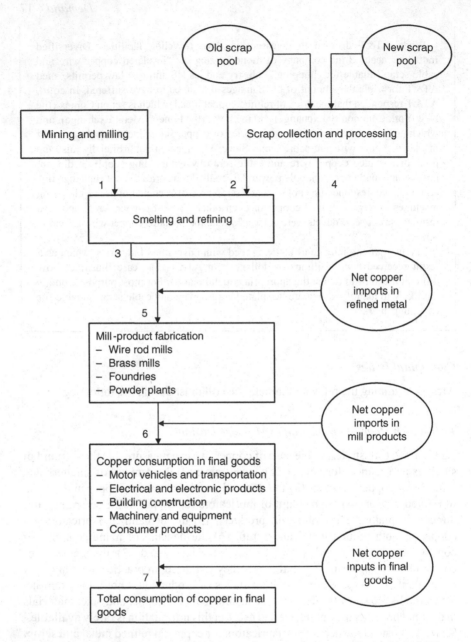

Figure 2.1 Copper consumption at various stages of use (source: Radetzki and Tilton (1990, Figure 2.1)).

Note: Flows at the points indicate copper demand on the part of: (1) smelters and refiners for copper in concentrates; (2) smelters and refiners for copper in scrap; (3) fabricators for copper in the form of refined metal; (4) fabricators for copper in the scrap they use directly in making ingot and mill products; (5) fabricators for copper in both refined metal and scrap; (6) manufacturers for copper in fabricated and semi-fabricated products; and (7) consumers for copper in the final goods they purchase.

2 In What Units to Measure Demand

Normally, we use tons or some other measure of weight (such as ounces in the case of silver and gold), again in part because the data are more readily available. Depending on the purpose of the analysis, however, dollars or some other measure of value may be more appropriate. Value reflects differences among commodities in their unit price and may capture as well improvements in quality over time.

3 Whether to Include All, Part, or None of Secondary Production

Secondary metal is produced by recycling scrap: old scrap that arises when products come to the end of their useful lives and new scrap produced in the process of making new products. Some non-metallic commodities such as gemstones and asbestos are also recycled. Just what secondary production should be considered in any particular study depends on the study's purpose. When trying to measure the amount of copper consumed by manufacturers in the final goods they sell to the public over a given period of time, a year for example, one normally wants to include secondary copper produced from old scrap but not secondary produced from new scrap. Including the latter entails double counting since the copper recycled as new scrap has already been counted when first used by fabricators. In practice, the available data often limit the possibilities for including or excluding secondary production.

4 How to Treat Changes in Inventories

Producers and consumers at each stage of production generally hold stocks of both their inputs and outputs. Depending on the purpose of the analysis it may or may not be appropriate to take account of changes in these stocks. Again, however, in practice, data on inventories are often not available, limiting the possibilities for adjusting demand for inventory fluctuations.

Determinants of Demand

An endless number of factors affect the demand for mineral commodities—new technology that reduces the fire hazard associated with the use of lithium in automobile batteries, increased funding for public transportation in London, permission to allow the construction of extensive hydroelectric facilities in the south of Chile, rising per capita incomes in India, the recovery of the Japanese economy. In analyzing demand, it is neither possible nor desirable to take account of all the possible determinants. There are simply too many. Moreover, the effects of most are so trivial they should be ignored in order to avoid needlessly complicating the analysis.

The art of good analysis—and it is an art, not a science—is deciding just which factors should and should not be considered. The answer is likely to vary not only with the commodity of interest, but also with the purpose and time horizon of the study. Technological change, for example, is not likely to alter greatly the demand for cobalt over the coming year, and so can probably be safely ignored when one is focusing on the short run. This is not the case when cobalt demand 20 or 30 years in the future is the issue.

The choice of which factors to consider and which to ignore is important, and will largely determine the usefulness of an analysis. For this reason, it is worthwhile to review the determinants often encountered in studies of demand.

1 Income or Economic Activity

Metals and other mineral commodities are used in the production of consumer and producer goods. So changes in the output of these goods have an immediate and direct impact on their demand. It is useful to distinguish between two types of changes in aggregate economic activity or income. The first are short-run changes that occur largely as a result of fluctuations in the business cycle. The second are longer-run changes driven by secular growth and structural change in the economy.

As income is among the most important variables affecting commodity demand, its influence is almost always taken into account. Many studies use GDP (gross domestic product) or industrial output for this purpose. More disaggregated measures may also be appropriate at times. For example, in assessing the demand for superalloys consumed in the production of jet engines, one might find aerospace production or even more specifically jet aircraft production a better measure of the relevant economic activity than GDP or even industrial production in general.

2 Own Price

Another important determinant of demand, which is also usually considered, is a mineral commodity's own price. For two reasons demand tends to fall with a rise in own price and vice versa. First, a higher price increases the production costs of the final goods in which metals and non-metallics are used. Assuming all or part of the higher costs are passed on to consumers, the price of the final goods will rise, causing the demand for both final goods and their inputs to fall. In many situations, however, this effect of a change in own price is modest. This is because the fraction of the total costs of final products accounted for by the individual metals or the non-metallics they contain is often quite small. A doubling or tripling of the price of lead or platinum, for example, increases the cost of producing automobiles by less than 1 percent.

Second, and generally more important, consuming firms may respond to higher inputs prices by substituting another material whose price has risen less or not at all. A rise in the price of aluminum, for example, is likely to encourage soft-drink producers to reduce their use of aluminum cans and increase their use of plastic and glass bottles. Material substitution of this nature, however, may take some time to effect. This is especially so where new equipment is necessary, personnel need to be retrained, and production techniques altered. As a result, a change in price for a mineral commodity may have only a modest effect on its demand in the short run and require a number of years to realize its full effect. Given its importance, we examine the nature of material substitution in greater depth later in this chapter.

3 Prices of Substitutes and Complements

The demand for a metal or non-metallic may be altered by prices other than its own. This is because most metals and non-metallics compete with other materials for markets. As a consequence, a change in the price of a competitive or substitute material can affect their demand. When the price of platinum rises, for example, the producers of fine jewelry tend to increase their demand for gold.

In some instances, a fall in the price of a metal or non-metallic may increase the demand for another commodity. In such cases, we say the two are complements. A lower price for iron, for instance, increases the demand for galvanized steel and in turn for zinc (a critical ingredient for galvanized steel). Similarly, a fall in the price of precious metals may stimulate the demand for gemstones.

As with changes in own price, changes in the price of substitutes or complements affect demand primarily by encouraging firms to alter production processes. Since this often takes some time, the impact of such price changes tends to be greater in the long run than in the short run.

4 Technological Change

New technology affects the demand for mineral commodities in several ways. First, it reduces the amount of material required in final goods. In the 1960s young men demonstrated their physical prowess by crushing metal beer cans in a single fist. Today, thanks to new paper-thin aluminum alloys, a six-year-old can do the same.

Second, new technology alters the ability of materials to compete for end-use markets. The aluminum can captured most of the beverage container market from the steel tinplate can in the 1970s thanks largely to the introduction of the two-piece aluminum can, new, thinner alloys, and other cost-reducing innovations. Another example is the use of lithium in rechargeable batteries where new technology made lead–acid batteries obsolete.

Third, new technology creates and destroys end-use markets. The rise of the computer and semiconductor industries, for instance, over the past 50 years generated major new uses for indium, silicon, germanium, and gallium arsenide. Should electric automobiles replace the internal combustion engine, the demand for platinum and palladium would suffer as the production of catalytic converters, their major market, collapsed. On the other hand, should lithium replace lead in new and more powerful automobile batteries, its demand would explode.

Measuring the impact of technological change is difficult, and as a result many studies simply ignore this determinant of demand. As noted earlier, this omission may be acceptable when the focus is on the short run. Over the longer run, however, it is harder to justify. Others have assumed that the impact of technological change on demand is highly correlated with time. When this assumption is valid, one can use a simple time trend to capture the effects of technological change (though it is important to remember that the time trend will also reflect the influence of the other excluded variables that rise or fall with time). When the impacts of innovation and technological change are random and discrete, there unfortunately is no good alternative other than assessing them explicitly and individually.

5 Consumer Preferences

Consumer preferences for the final goods and services that people purchase shape the output of the end-use markets consuming metals and other mineral commodities. When preferences shift toward buying books and other goods online rather than in retail stores, the demand for stores and all the materials used in their construction tends to decline.

Consumer preferences vary over time and among countries for many reasons. It is no secret that higher gasoline prices stimulate the demand for hybrid and all-electric automobiles. Demographics and the age distribution of a country's population may also be important. Brazil and other developing countries have a larger share of their population than developed countries between the ages of 18 and 35, a time when many individuals are engaged in creating new families and spending a relatively large portion of their income on housing, automobiles, refrigerators, and other, material-intensive consumer durables. Cultural differences, too, can be important. They help explain, for example, why India is the world's largest consumer of gold jewelry.

New technology may also alter consumer preferences by creating new and better products. The rapid growth of the airline industry during the latter half of the twentieth century increased the demand for aluminum and titanium, while reducing the need for steel in railroad passenger cars and ocean liners. Another, more recent, example is the worldwide explosion in the demand for personal computers, tablets, and smartphones.

Consumer preferences can also change simply in response to shifts in personal tastes. In some instances, these shifts are driven by advertising and psychological considerations that are not fully understood.

6 Government Activities

Government policies, regulations, and actions also influence the demand for mineral commodities. This is probably most dramatically illustrated when government policies lead to war and the production of armaments and other military equipment. In peacetime, changes in government expenditures on education, defense, highways, and other public goods alter the mix of final goods and services produced and consumed. Fiscal, monetary, and social welfare policies change the distribution of income, the level of investment, and the rate of economic growth. In recent years, policies encouraging urbanization in China and elsewhere have stimulated the use of copper, steel, sand and gravel, and a host of other materials.

On the other hand, worker health and safety legislation, environmental standards, and other regulations can restrict the use of certain minerals in particular applications. This is nicely illustrated by the demand for lead, where over the past several decades regulations designed to protect public health have proscribed the use of lead in paints and gasoline and largely curtailed its use in solders. As a result, the demand for lead, other than for automobile batteries, has declined considerably over the past 50 years in the United States and the world as a whole. A similar story can be told for asbestos.

The Demand Function and Curve

The *demand function* reflects the relationship between the demand for a mineral commodity (or any other good) and its major determinants, such as those just identified. It is often expressed mathematically. In some studies, for example, the demand for a particular commodity, such as aluminum, during a given year t (Q_t^d) is assumed to depend on three determinants—GDP during the year (Y_t), average own price during the year (P_t^o), and the average price of copper or another important substitute during the year (P_t^s):

$$Q_t^d = f(Y_t, P_t^o, P_t^s) \tag{2.1}$$

Several attributes of Equation 2.1 are worth noting. First, it assumes there are only three important determinants of demand. In many situations, other relevant variables belong in the demand function as well.

In addition, it considers only the short-run effects of changes in the determinants of demand, since demand this year depends solely on this year's income and prices. For income or GDP, this may not be a significant limitation since the demand for metals and non-metallics tends to respond rapidly to changes in economic activity. As we have seen, this is much less the case for changes in prices. Producers take time to substitute one material for another and in other ways to respond fully to changes in own price or the prices of substitutes. This means that the demand this year depends not just on prices this year, but also on prices a year ago, two years ago, and perhaps even further in the past.

Finally, Equation 2.1 identifies the presumed important determinants of demand. But it does not indicate the exact nature of the relationship between demand and its determinants. Most studies assume a simple linear or logarithmic relationship, similar to those shown in Equations 2.2 and 2.3. This is in part because such relationships are easy to estimate using well-known econometric software packages. In addition, with the logarithmic relationship shown in Equation 2.3, the coefficients provide direct estimates of the demand elasticities (discussed in the next subsection).[3] The linear and logarithmic relationships, however, entail rather strong assumptions about the nature of the demand function, whose validity is often difficult to assess. Fortunately, when the evidence suggests that other relationships are more appropriate, modern computational power and software allow their estimation even when they are complex and non-linear.

$$Q_t^d = \beta_0 + \beta_1 Y_t + \beta_2 P_t^o + \beta_3 P_t^s \tag{2.2}$$

$$\ln Q_t^d = \mu_0 + \mu_1 \ln Y_t + \mu_2 \ln P_t^o + \mu_3 \ln P_t^s \tag{2.3}$$

The *demand curve* is derived from the demand function. It shows the demand for a mineral commodity (that is, how much can be sold) at various own prices over a year or some other time interval on the assumption that income, the prices of substitutes, and other determinants of demand remain fixed at current levels or some other designated levels. Demand curves are often drawn with a negative or downward slope, like those shown in Figures 2.2 and 2.3, since typically demand falls when price rises. There are, though, exceptions. When buyers want a particular amount of a product, no more and no less, regardless of price, the demand curve is vertical. When buyers demand none of a commodity above a certain price and an insatiable amount at that price, the curve is horizontal. The demand curve can even be upward sloping, when a rise in price causes buyers to want more of a commodity.

Several other characteristics of demand curves should also be noted:

1   A change in a commodity's own price causes a movement along the demand curve. A change in any of the other determinants of demand found in the demand function causes a shift in the curve itself. As a result, in Figure 2.2, demand can increase from Q_1 to Q_2 because its price falls from P_1 to P_2, causing a movement down the curve. Alternatively, a shift in the curve from D_1 to D_2, caused for instance by a rise in GDP, can produce the same increase in demand without any change in the commodity's price. It is important to keep this distinction between a movement along the demand curve and a shift in the curve clear.

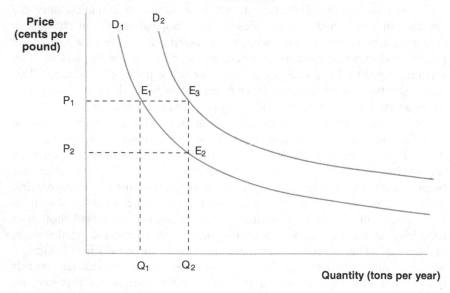

Figure 2.2 Movements along and shifts in the demand curve.

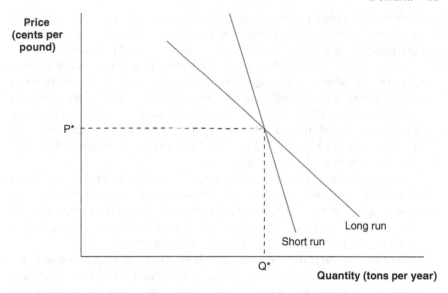

Figure 2.3 Demand curves in the short and long run.

2   The same commodity may have many different demand curves. As noted earlier, one can measure demand for the entire market, which reflects the total quantities all buyers are willing to purchase at various price levels. Alternatively, on the buyers' side of the market, we can construct demand curves for individual buyers, for regional or national markets, for country imports, for particular consuming sectors or industries. We can also distinguish between the demand curve for consumers and the demand curve for speculators and stockpiles. On the sellers' side of the market, a similar breakdown is possible. Normally, though, it is the market demand curve that is of primary interest as it, along with the market supply curve, determines price.

3   The market demand indicated by any point on the demand curve is not necessarily the same as consumption or production (even though the horizontal axis for many demand curves is identified as output). For example, when the US government is buying tin for its strategic stockpile or when speculators or others are accumulating tin for their inventories, demand and production are greater than consumption. Alternatively, if the government and speculators are drawing down their stocks, demand and production will be less than consumption. On the other hand, when producers are liquidating their inventories, production is less than demand and consumption.

Over the longer run, a decade for example, the differences between consumption, production, and market demand are small and usually can be ignored. This is in part because inventory changes over the longer term tend to cancel out and in part because any net changes in inventories will be small compared to cumulative consumption and production over longer periods. Often, however, demand curves indicate how much buyers will purchase at various prices over a year or even a shorter period. In such cases, changes in the stocks held by producers, consumers, and third parties can cause significant discrepancies among market demand, consumption, and production.

4   The demand curve shows how demand responds to a change in price over a given period of time. In this regard, economists typically distinguish between the short run (a period sufficient for firms to adjust output by altering their variable inputs, such as their labor, energy, and materials) and the long run (a period long enough for firms also to vary their plant and equipment and other fixed inputs). As Figure 2.3 shows, the responsiveness of demand to price tends to be greater in the long run.

Just how long are the short and long runs? In practice, this varies with the commodity and other considerations, but as a rough rule of thumb we often assume the short run is a period of up to five years and the long run a period of five years or more. While there are exceptions, these are reasonable estimates for many metallic and non-metallic industries.[4]

The adjustment periods associated with the short and long runs should not be confused with the time interval over which the demand curve measures demand. For example, Figures 2.2 and 2.3 measure demand in terms of tons per year. The long-run demand curve does not measure how much of a commodity will be demanded over the long run—for example, over the next five to ten years. Rather it indicates how much will be demanded per year five to ten years hence, assuming price and the other determinants of demand remain at their indicated values.

Of course, price and the other determinants of demand will not remain unchanged over this period. As a result, it is misleading to think of the long-run demand curve as indicating what demand will be five to ten years hence. Better to think of it as showing the equilibrium towards which demand is moving today over the long run, given today's price and other current conditions. Long before this equilibrium is reached, price and other determinants will change, causing the trend in demand to alter its course and follow a new path towards a different equilibrium.

5   The downward-sloping demand curve, as commonly drawn, also assumes that the relationship between price and demand is continuous, reversible, and deterministic. Continuity implies that one can draw the curve without picking up the pen. So the curve, like those shown in Figures 2.2 and 2.3, has no jumps or breaks. This assumption is not valid when there are major changes—upward or downward—in demand as price varies. For example, for those metals and other mineral commodities whose consumption is concentrated in a few major end uses, price may rise over a wide range with little or no effect on demand. Then, at some threshold, an alternative material becomes more cost effective in an important application, causing a sharp drop in demand. Such discontinuities may be found in both short- and long-run demand curves. When they occur, they can have a substantial impact on demand.

Reversibility means that demand will return to its prior level if price, after a rise or fall, returns as well to its prior level. In short, one can move up and down the demand curve without causing the curve itself to shift. Reversibility may hold for short-run demand curves, but this is less likely for long-run curves. Over the long run, price changes are more likely to foster material substitutions that entail the purchase of new equipment, the retraining of employees, loss of production, and other conversion expenses. Once a firm encounters these costs, it will be hesitant to switch back to the material it was previously using, even if price falls back to the status quo ante.

Deterministic implies there is no uncertainty. Given the curve and a price, the quantity demand is known for sure. Where uncertainty exists, we say the relationship between demand and price, as well as the other determinants of demand, is stochastic. For this reason, another term is often added to Equations 2.1–2.3, known as the error term or the disturbance term. It captures the uncertainty, when there is uncertainty, in these relationships.

Elasticities

In assessing commodity markets, mineral economists and others often need to know how sensitive demand is to changes in prices and income. For example, if a mine accident in Russia were to reduce world nickel output by 1 percent, one might like to know how much higher nickel prices would be over the coming year. If nickel demand is highly sensitive or responsive to price, the increase would be modest. However, if demand is little affected by price changes, as is often the case in the short run, the price increase required to curtail demand to the lower, available supply could be quite large.

The *elasticity of demand with respect to price*, or simply the price elasticity of demand, is the measure mineral economists normally use to assess how demand responds to changes in a commodity's own price. As Equation 2.4a shows, the price elasticity of demand is the partial derivative of demand with respect to price times the ratio of price to demand. Most of us find it easier to remember the price elasticity of demand as simply the percentage change in demand caused by a 1 percent change in price, as shown in Equation 2.4b. Since a rise in price normally causes a fall in demand and vice versa, we multiple the resulting change in demand by –1, so that the elasticity itself is a positive number. If a 1 percent price change causes demand to change by more than 1 percent, we say that demand is elastic. If the change is less than 1 percent, demand is inelastic.

$$E_{Q_t^d, P_t^o} = -\frac{\partial Q_t^d}{\partial P_t^o} \cdot \frac{P_t^o}{Q_t^d} \tag{2.4a}$$

$$= -\frac{\text{percent change in } Q_t^d}{\text{percent change in } P_t^o} \tag{2.4b}$$

Because the derivative of demand with respect to price is equal to the inverse of the slope of the demand curve, where two curves cross, the price elasticity of demand will be lower for the curve with the steeper slope. In Figure 2.3, this means that demand at the point where the curves intersect is more elastic for the long-run curve. This is consistent with what we would expect, as consumers have more opportunities to increase or decrease their use of a material in response to a price change, the longer the time period they have to adjust.

If the relation between demand and price is linear, as Figure 2.3 and Equation 2.2 assume, the slope of the demand curve is everywhere the same. This means that the price elasticity of demand decreases as one moves down the demand curve and the ratio of price to demand falls. Consequently, other than their point of intersection, we should be careful in comparing the demand elasticities of two curves. The steeper curve will not necessarily have the lower elasticity everywhere.

When the relation between demand and price is logarithmic, as assumed in Equation 2.3, the price elasticity of demand is the same everywhere. So one can easily compare the elasticities of two curves, even at points where they do not intersect. For this reason, analysts often assume a logarithmic relationship between demand and price. As noted earlier, however, this assumption is appropriate only if there are good reasons to believe the true relationship between the two is indeed logarithmic.

So far we have considered only the price elasticity of demand. Separate elasticities can be defined for all the variables affecting demand, though in practice we normally encounter only two others—the elasticity of demand with respect to the price of a substitute (or complement) and the elasticity of demand with respect to income.

The elasticity of demand with respect to the price of a substitute, often called the cross (price) elasticity of demand, measures the percentage change in the demand for a material such as copper caused by a 1 percent change in the price of a substitute such as aluminum or plastics. It, too, will normally increase as we move from the short to the long run, since the opportunities to take advantage of a change in a substitute's price grow with the adjustment period.

The elasticity of demand with respect to income, or more simply the income elasticity of demand, in a similar manner measures the percentage increase in demand caused by a 1 percent rise in GDP or some other measure of income. Here again we expect to find differences in the elasticity, depending on the adjustment period, though the differences are not the same as for own and cross price elasticities.

In the short run, a period more-or-less up to five years, the business cycle largely drives changes in GDP and the macro economy. Four economic sectors—capital equipment, construction, transportation, and consumer durables—consume the lion's share of most metals and other materials, and use them to build automobiles, washing machines, houses and office buildings, new machinery, and other such items. When the economy is booming, these end-use sectors tend to expand even more rapidly. They also suffer more severely when the economy is in a recession. Because fluctuations in GDP over the business cycle cause even larger fluctuations in the output of these four end-use sectors and in turn in the demand for materials, normally we expect the income elasticity of demand to be greater than 1—or elastic—in the short run.[5]

In the long run, GDP changes as a result of secular economic growth. The traditional and still quite common presumption is that metal and material demand tends to expand more or less at the same pace as the secular growth of the economy. When this is the case, the income elasticity of demand in the long run is unity or close to unity. In the 1970s and 1980s, however, proponents of the intensity-of-use hypothesis (discussed below) and others challenged this presumption. As discussed in the next section, there are now reasons to believe that the income elasticity of demand in the long run is less than 1 in developed countries and more than 1 in developing countries.

Intensity-of-Use

The use of a commodity, such as copper, depends on the amount consumed in individual consumer and producer goods, such as automobiles, houses, and electric power generators, and the total output of each of these goods. This relationship, which is an identity, is shown in Equation 2.5. Q_t^d is the demand (strictly the consumption) of copper, O_{it} is the output in physical units of the ith final product, a_{it} the amount of copper consumed per unit in producing the ith final product, and n_t the number of final goods, all in year or period t.

$$Q_t^d = \sum_{i=1}^{n_t} a_{it} O_{it} \tag{2.5}$$

If we define the ratio of the output of the ith good (O_{it}) to national income (Y_t) in period t as b_{it} (= O_{it}/Y_t) and substitute for O_{it} in Equation 2.5, we obtain:

$$Q_t^d = Y_t \sum_{i=1}^{n_t} a_{it} b_{it} \tag{2.6}$$

This identity shows that commodity demand can vary for a country or the world as a whole over time (or among countries at any point in time) as a result of changes in: (a) the level of income (Y_t), (b) the material composition of products (a_{it}), which reflects the mix of materials used to produce individual products, and (c) the product composition of income (b_{it}), which reflects the mix of goods produced by the economy.

If we now define intensity-of-use (IU_t) in the normal manner as the ratio of a commodity's consumption (Q_t^d) to income (Y_t), it is clear that intensity-of-use is simply a function of the product composition of income and the material composition of products, as shown in Equation 2.7.

$$IU_t = Q_t^d \Big/ Y_t = \sum_{i=1}^{n_t} a_{it} b_{it} \tag{2.7}$$

The inverse of intensity-of-use, it is worth highlighting, is simply productivity. When intensity-of-use is falling, the income or output realized per ton of copper consumed is rising. This suggests that intensity-of-use, like aggregate measures of labor or capital productivity, is itself of some intrinsic interest. It also makes clear that declining intensity-of-use, though often lamented by producing firms and countries, is not necessarily bad. Indeed, from the point of view of resource efficiency and the long-run availability of mineral commodities, quite the contrary is true.

Intensity-of-Use Technique

The intensity-of-use technique is a simple procedure, which on occasions is used to assess changes over time (or differences among countries at a given point in time) in commodity demand. It is based on the identities shown in Equations 2.5, 2.6, and 2.7. The first step entails measuring how much of a rise or fall in the demand is due to changes in income and how much to changes in intensity-of-use. The second step assesses the contribution of changes in the material composition of products and in the product composition of income in explaining changes in intensity-of-use. The third step attempts to explain changes in the product composition of income, by looking at inter-sectoral shifts in the economy (such as the growing importance of the service sector) and at intra-sectoral shifts (such as the increase in electronic and other high-technology products in the manufacturing sector). Changing consumer preferences are largely behind such shifts. Finally, the fourth step focuses on changes in the commodity's own price, the prices of substitutes, and the other fundamental determinants of demand altering the material composition of products.

When changes in income are largely responsible for shifts in demand, as is often the case over the short run, then a good understanding of macroeconomics and of the forces driving GDP up or down is needed to explain demand trends. When shifts in the product composition of income are most important, a good knowledge of how and why consumer preferences are evolving is required. Finally, when changes in the material composition of products are largely driving demand trends, one needs to focus on technological change and changing prices for the materials of interest and their substitutes.

The Intensity-of-Use Hypothesis

The intensity-of-use hypothesis should not be confused with the intensity-of-use technique. It is a true hypothesis in the sense that it can be tested and rejected, not simply a procedure based on a set of irrefutable identities. Specifically, it postulates that the intensity-of-use of a commodity varies with economic development and per capita income. At early stages of economic growth, a country's material requirements are quite low as its economy consists largely of unmechanized, subsistence agriculture. As industrialization takes place, manufacturing, construction, and other material-intensive activities expand, causing intensity-of-use to rise. As development continues, the need for houses, factories, roads, automobiles, and communication towers is eventually satisfied, and demand shifts toward education, medical care, and other services. Since the service sector presumably is less material-intensive than manufacturing and construction, this shift in consumer preferences leads first to a slowing and then to a reversal in the upward rise in intensity-of-use as development and per capita income advance.

In the 1970s and 1980s a number of organizations and analysts used the intensity-of-use hypothesis to forecast the demand for a number of mineral commodities.[6] The big advantage of this approach is its simplicity. For example, as Figure 2.4 shows, one can estimate the inverted U-shaped relationship between per capita income and intensity of zinc use in the United States on the basis of historical data. All that is still needed to predict the US demand in, for example, 2050, are forecasts of the country's GDP and population for that year. These are readily available from a number of sources.

As an illustration, assume the population of the United States in 2050 will be 400 million people and its GDP 30 trillion dollars (reflecting a growth rate in real GDP of around 2 percent per year). Then income per capita will be 75,000 dollars. At this per capita income, Figure 2.4 anticipates an intensity of zinc use of approximately 50 tons for every billion dollars of GDP. Multiplying this figure by the country's estimated GDP in 2050 produces a forecast of 1.5 million tons for annual zinc demand in the United States by 2050.

This forecasting technique does suffer from some shortcomings. The most serious is the implicit assumption that intensity-of-use depends solely on per capita income, and hence that the inverted U-shaped curve shown in Figure 2.4 is stable. While inter-sectoral shifts—first, from agriculture to manufacturing and construction and then eventually to services—presumably do cause intensity-of-use to vary with economic development, we know that changes in the material composition of products do so as well.

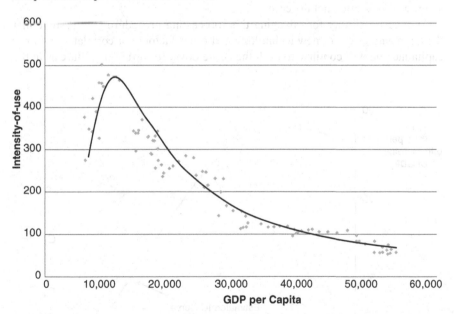

Figure 2.4 Intensity of zinc use in the United States, 1929–2014 (sources: US Census Bureau (2014) for population; US Geological Survey (annual) for US slab zinc apparent consumption; US Bureau of Economic Analysis (2014) for real US GDP).

Notes:
1 Intensity-of-use is measured in tons per billion constant (2014) dollars of GDP. Per capita income is GDP, again measured in constant (2014) dollars, divided by population.
2 The line is a freehand-drawn curve.

As a result, material price changes, new technologies, and other factors whose influence is unlikely to vary closely with per capita income also affect intensity-of-use. This means that the true inverted U-shaped relationship between intensity-of-use and per capita income shifts over time. As shown in Figure 2.5, this curve often shifts downward thanks to new resource-saving technology. There are, however, exceptions. Both material substitution and new technology occasionally push the intensity-of-use curve upward. The development of the lithium battery, for example, increased the intensity of lithium use in the United States and many other countries.

When the true intensity-of-use curve shifts, the observed historical data used to estimate the curve come from different curves. As a result, the estimated curve (i.e., the solid curve in Figure 2.5 and the curve shown in Figure 2.4 for zinc) is a hybrid composed of various points on different true curves and is unlikely to provide a good approximation for any of the true curves.

If the net influence of those factors not closely correlated with per capita income were quite modest, the shifts in the intensity-of-use curve over time would be small and the forecasts of future demand obtained in this manner would still be reliable. Unfortunately, the available evidence, particularly regarding the importance of technological change on demand, discussed in the next section, suggests this is often not the case.

Another possibility for salvaging this forecasting procedure is to assume that the aggregate effect of new technology and other factors not correlated with per capita income will continue to push the curve down (or up) in the future as in the

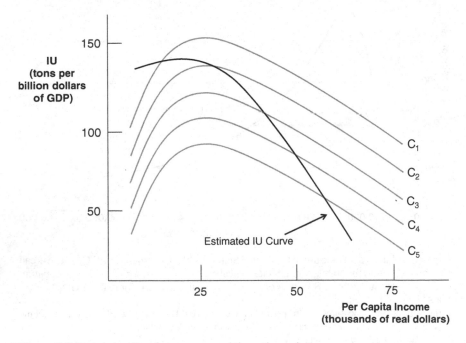

Figure 2.5 True intensity-of-use curves and the estimated curve.

past. When this is the case, the estimated intensity-of-use curve may capture both the influence of those determinants of demand correlated with per capita income and those responsible for the shifts in the curve over time, and so provide reasonably good forecasts. However, this approach guarantees poor forecasts just when they are needed most, that is, when significant breaks with the past are occurring.

For these reasons, the intensity-of-use hypothesis is not widely used for forecasting today. A more common approach, instead, is to identify the major end uses (both actual and potential) for a mineral commodity, then use various methods to forecast the demand for these important end uses, and finally to assess future changes in the material composition for each of these end-use products.

It would be wrong, however, to dismiss the intensity-of-use hypothesis as of little use. For many metals and other commodities intensity-of-use does first rise and then fall with per capita income. So the inverted U-shaped intensity-of-use curves shown in Figure 2.5 do exist. While other determinants of demand cause them to shift over time, reducing their usefulness for forecasting, they still can provide useful insights into the nature of commodity demand.

Of particular importance for our purposes, the intensity-of-use hypothesis implies that a 1 percent increase in income in developing countries that causes per capita income to rise will generate more than a 1 percent increase in demand, as otherwise rising per capita income would not increase the intensity-of-use. This means in developing countries the long-run income elasticity of demand for metals and other mineral commodities is typically greater than 1. On the other hand, in developed countries where an increase in per capita income is associated with a decline in intensity-of-use, we expect the long-run income elasticity of demand to be less than 1.

Material Substitution

As the intensity-of-use technique shows, in the first instance material demand trends can be explained by changes in income, in the product composition of income, and in the material composition of products. Since material substitution—that is, the replacement of one material for another in the production of particular goods—is largely responsible for changes over time in the material composition of products, this section explores in more depth the nature of material substitution.

In analyzing material substitution and the forces behind this important activity, mineral economists have traditionally emphasized the role of relative material prices, suggesting that material substitution should be particularly important in end-use applications where the prices of competing (and complementary) materials are experiencing substantial changes. There is, however, an alternative view that argues that technological change is the dominant force behind material substitution, and as a result material substitution is most likely to occur in end-use industries whose production technologies are evolving rapidly even if relative material prices are changing little.

The Traditional View

The traditional view considers the substitution of one material for another to be simply part of the broader, more general process by which firms determine the particular inputs or factors of production they use. According to the theory of the firm, that branch of microeconomics concerned with this process, the potential for one material to substitute for another—or more generally for one input, such as capital, to substitute for another, such as labor—is captured by what is called an *isoquant curve*.

This curve shows the various combinations of any two inputs required to make a given amount of a particular good. For example, the isoquant curve Q_1 shown in Figure 2.6a might reflect the various tonnages of steel and aluminum needed to produce 60,000 light trucks of a particular model. Normally, as shown in this figure, an isoquant curve has a negative slope, since reducing the amount of one material (steel) increases the amount needed of the other (aluminum) to maintain production at a particular level.

Also, as shown in Figure 2.6a, an isoquant curve is typically drawn convex to the origin, reflecting the presumption that as more of one input is used in place of the other, greater quantities of the first are needed to replace one unit of the second. In our light truck example, this assumption implies that the first substitutions of aluminum for steel involve trim or other applications where a little aluminum can replace a pound of steel. As this substitution continues, however, steel parts that carry heavier loads and rely on greater strength have to be replaced, and this requires much more aluminum for each pound of steel replaced.

For every level of output for light trucks there exists a corresponding isoquant curve indicating the various quantities of steel and aluminum required. The isoquant curve Q_2 in Figure 2.6a, for example, might reflect the steel and aluminum needed for 50,000 light trucks.

After the firm determines its desired level of output and hence the particular isoquant curve it wants to be on, it must select the mix of steel and aluminum it will use. This involves picking the point on the relevant isoquant curve that minimizes costs. To determine this point, economic theory constructs what it calls *isocost curves*. The straight lines C_1 and C_2 in Figure 2.6a are examples. They show the various combinations of two factors—steel and aluminum—that can be acquired for a given amount of money, for example 50 million dollars. For this sum, if the price of aluminum is 2,500 dollars per ton and steel 1,000 dollars per ton, one can buy 20,000 tons of aluminum and no steel, 50,000 tons of steel and no aluminum, or various combinations of both, such as 10,000 tons of aluminum and 25,000 tons of steel. The isocost curve C_1 shows all of these possible options. It intersects the steel (vertical) axis at 50,000 tons and the aluminum (horizontal) axis at 20,000 tons, and has a slope of -2.5 reflecting the negative of the ratio of the price of aluminum to the price of steel.

A similar isocost curve (C_2) can be constructed for a total expenditure of 40 million dollars or for any other amount. Isocost curves associated with total expenditures of less than 50 million dollars, such as the curve C_2, will lie between

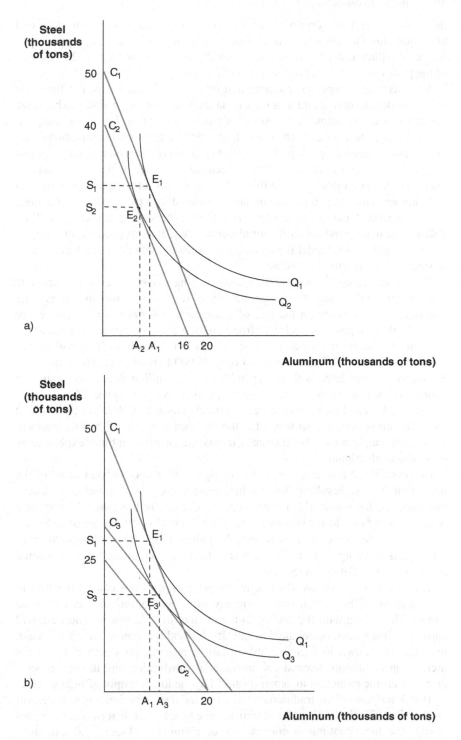

Figure 2.6 Material demand determined by conventional isoquant and isocost curves with: (a) constant prices; and (b) changing prices.

the curve C_1 and the origin, since they cannot purchase as much steel and aluminum. Just the opposite is the case for isocost curves representing expenditures above 50 million dollars. All these curves will have the same slope (–2.5) as long as the prices of steel and aluminum do not change.

To determine its optimal consumption of steel and aluminum, the firm first selects its desired output and thus the isoquant curve on which it wants to be. If we assume its desired output is 60,000 light trucks, it needs to be on the isoquant curve Q_1 in Figure 2.6a. It then selects the particular point on that curve that minimizes its input costs. This is the point (E_1) where the isoquant curve is just tangent to the isocost curve C_1. This is because any other point on the isoquant curve will lie on an isocost curve that reflects a greater total expenditure on steel and aluminum than the 50 million dollars associated with the curve C_1. Of course, there are isocost curves associated with total expenditures under 50 million dollars, but none provides sufficient aluminum and steel to produce the desired output of light trucks. Under these conditions, the optimal mix of inputs is S_1 tons of steel and A_1 tons of aluminum.

We are now ready to consider the question: What drives material substitution? Why might the firm want to change the mix of steel and aluminum it uses? The traditional view focuses on the role of changes in material prices. Assume, for example, that the price of steel rises from 1,000 to 2,000 dollars per ton while the price of aluminum remains at 2,500 dollars per ton. Now 50 million dollars will still buy 20,000 tons of aluminum, but only 25,000 tons of steel. This causes the isocost curve associated with an expenditure of 50 million dollars to rotate in a counter-clockwise manner. The new curve, shown as C_2 in Figure 2.6b, now has a slope of –1.25 and intersects the steel (vertical) axis at 25,000 rather than 50,000 tons. This curve is no longer tangent to the isoquant curve Q_1. If the firm wants to maintain its output at 60,000 light trucks, it must increase its combined expenditures on steel and aluminum.

However, if the firm raises its price for light trucks to cover part or all of the increase in its costs resulting from the higher steel price, the demand for its trucks may fall, causing the firm's desired output to decline. In this case, the firm may want to operate on a lower isoquant curve, that is one closer to the origin in Figure 2.6b, such as the curve Q_3. In this case, the point (E_3) on the new isoquant curve (Q_3) that is just tangent to an isocost curve (C_3) will determine the new optimal mix of steel and aluminum $(S_3$ and $A_3)$.

The new mix shows that the rise in the price of steel has caused the firm to replace some of the steel it was previously using with aluminum, as we would expect. This, along with the decline in the number of light trucks produced, will cause the firm's demand for steel to fall. Its demand for aluminum, on the hand, may rise (as shown in Figure 2.6b) or fall, depending on whether or not the increase in aluminum demand due to material substitution and the higher steel price offsets the reduction in demand caused by the lower output of light trucks.

This description of the traditional view of material substitution is to some extent an oversimplification. Mineral economists have long known that isoquants are not always the nice continuous convex curves pictured in Figure 2.6 and most introductory microeconomic textbooks. Technology may require that inputs be used in fixed proportions, especially in the short run, when it is often difficult to modify or replace existing equipment. In such cases, a given output of a final good requires one specific combination of inputs, and the isoquant collapses to a single point, as shown in Figure 2.7a. Alternatively, there may be two or three mutually exclusive production processes that a firm may use to produce a given output, each requiring fixed but different input proportions. In these circumstances, the isoquant becomes two or three points.

Even where a continuous isoquant exists, it may not be convex. Aluminum and plastic are both used as siding in the construction of houses in the United States. If the price of plastic falls, there is no reason to believe that the amount of plastic required per pound of aluminum has to increase as the use of plastic increases. The isoquant curve in this case is linear, as shown in Figure 2.7b, rather than convex, and the optimal mix of inputs is usually found at one end or the other of the linear isoquant curve.

It is also recognized that existing technology determines the shape and location of the isoquant curve. As a result, the creation and diffusion of new technology can cause these curves to rotate or shift inward over time, altering the mix of inputs and fostering material substitution in the process.

The Alternative View

While the traditional view of material substitution recognizes that isoquant curves may possess discrete breaks, may be linear rather than convex, and may shift over time as technology changes, these possibilities are not emphasized. They are raised more as afterthoughts or as caveats and exceptions that in some instances may complicate the analysis. The role of technological changes is largely ignored, and in this sense the traditional view is static.

The alternative view, by contrast, emphasizes technological change and downplays material prices. In most cases, it maintains that the driving force behind material substitution is a shift in the isoquant curve caused by the introduction of new technology, rather than a shift in the isocost curve caused by a change in relative material prices.

New technology, according to the alternative view, is more important than material prices for two reasons. First, in many applications material prices can vary over a broad range without producing any movement on the isoquant curve and thus any change in the mix of materials. This is the case, for example, whenever the relevant isoquant is not a nice continuous convex curve, but is instead a point, a series of points, a straight line, or a series of linear segments. Even with a continuous convex curve, changes in material prices may not produce a movement along the curve if production is currently taking place at one end or the other.

Second, the past 50 years have witnessed tremendous advances in material technology. New metal alloys, polymers, ceramics, and composites now compete with the more traditional metals and materials. New innovations have

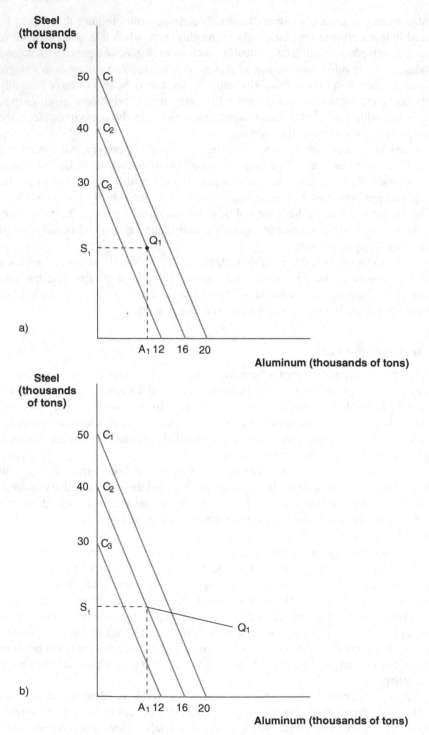

Figure 2.7 Material demand determined by point and linear isoquants: (a) point isoquant; and (b) linear isoquant.

simultaneously enhanced the strength, corrosion resistance, and other physical properties of steel and the other traditional materials, allowing them to continue to compete in many end-use markets. As a result, design engineers in all material-consuming industries—construction, automobile, aerospace, packaging, consumer durables, and capital equipment—enjoy a choice of materials that their predecessors several decades ago would greatly envy. Every year this choice becomes richer and more extensive.

The traditional and alternative views of material substitution, it is important to note, are not mutually exclusive. The real world is sufficiently rich and complex that one can find material substitution driven entirely by prices changes, entirely by technological change, and by both. Indeed, there are instances when neither is important, where material substitution is the result of changes in government policies or consumer preferences (see Box 2.2).

Box 2.2 The Use of Lead in Paint

For decades many paints contained lead as a pigment. Lead accelerates the drying process, enhances durability, and reduces corrosion by resisting moisture. Lead is, however, a toxic substance. It causes damage to the nervous system, impairs hearing, retards mental and physical development, and in large enough dosages can be fatal. It is particularly dangerous for young children. In 1978 the US government banned paints containing more than 0.06 percent lead from residential use. As a result, paint producers substituted titanium oxide, considered safe enough to use in food coloring and toothpaste, for the lead compounds previously used.

Nevertheless, it is still worthwhile to inquire whether changes in material prices or changes in technology are most often responsible for material substitution. Although the available studies are limited and largely confined to a few case studies, such as the US beverage container industry (see Box 2.3), they do suggest new technology and shifts in the production isoquants are often more important than changes in material prices and movements along stationary isoquants (Tilton 1983; 1991).

Box 2.3 Material Substitution in the US Packaged Beer Container Market

One interesting example of the available case studies of material substitution— conducted by Frederick Demler in the late 1970s as part of the research for his PhD dissertation (Demler 1980, 1983)—focuses on the US packaged beer container market.

In 1950, as Figure 2.8 shows, packaged beer (as opposed to beer in barrels) was shipped in three types of container. The returnable bottle was by far the most important, claiming some 70 percent of the market. The share of the one-way bottle was about 5 percent, with the three-piece tinplate can accounting for the rest of the

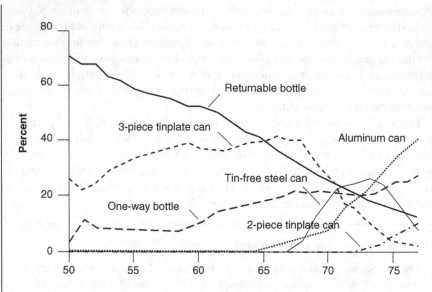

Figure 2.8 Percentage of packaged beer shipments by container type, 1950–1977.

market. The three-piece tinplate can was largely steel with a coating of tin, a bit of chrome, and often an aluminum top (for easy opening). As its name indicates, it was constructed from three separate pieces—a circular top and bottom and a rectangular piece that was rolled into a cylinder.

By 1977, the last year of data considered by Demler, the aluminum can enjoyed the largest share of the market. First introduced in 1958, it took off several years later when Reynolds Aluminum produced the first two-piece aluminum can, whose bottom and sides are constructed from one piece of aluminum.

The returnable bottle, the three-piece tinplate can, and the tin-free steel can (introduced in 1967) were headed for oblivion. Meanwhile, the one-way bottle was slowly increasing its market penetration, along with the two-piece tinplate can (introduced in 1971). In his thesis Demler shows how these changes in market share affected the consumption of the various materials—tin, steel, chrome, aluminum, and glass—used in beer containers. He then examined how the relative prices of these materials evolved over the 1950–1977 period and found little correlation between the changes in a material's price and its market share. Rather, the evidence showed that market shares changed largely in response to the new technological developments and innovations that producers generated in their competitive efforts to capture more of the packaged beer container market.

Demler also examined over the same period the material consumption of the US soft drink container industry and again found that material substitution was driven largely by innovation and new technology rather than changes in relative material prices.

Here, it is also worth noting that the contribution of material prices and technological change to material substitution cannot always be cleanly separated. Changes in material prices may themselves be the result of new innovations in mining or metallurgy. Alternatively, changes in material prices often spur the research and development efforts that lead to new and better materials and in turn material substitution. This means that changes in material prices and technology have both a direct effect on demand and an indirect effect, realized by altering one or more of the other determinants of demand.

This raises an intriguing issue. When one estimates the elasticity of demand for a commodity with respect to a change in its price or the price of a close substitute, does one include both the direct and indirect effects (where the latter includes the effect of the new technology that price changes induce)? While the answer to this question can vary with the purpose of the analysis, in many instances one is interested in the total effect of a price change. In such cases, the indirect effect should be taken into account, making the long-run elasticity larger—and often much larger—than would otherwise be the case. Including the indirect effect also means the demand curve is no longer reversible. The reduction in demand caused by a rise in its price is in part the result of price-induced new technology, which is unlikely to be abandoned or exactly reversed should price return to its original level.

Given the importance of the indirect effect of material prices arising from their influence on the rate and direction of technological change, our understanding of the relationship between material prices and demand, at least as conventionally portrayed in the theory of the firm and economics in general may be flawed. This relationship is described by the downward-sloping demand curve, which is normally drawn as continuous and reversible. However, if material prices primarily affect material demand indirectly by shaping the level and direction of R&D, and if material substitution in turn is an important determinant of demand, these assumptions are no longer likely to hold. A rise in price over a limited range may have little or no effect on demand, but once beyond some threshold it may stimulate new technology that shifts the demand curve downward. After the new technology is introduced, it is unlikely the process can be reversed. A fall in price to its original level will not recapture the lost demand.

Perhaps of even greater importance, the traditional view of material substitution may impede efforts to learn more about this important economic activity. In examining material substitution, as noted earlier, it encourages us to concentrate on changes in material prices and the factors responsible for these changes. It predicts material substitution will occur mostly in those uses experiencing significant changes in the prices of competing materials. The alternative view, on the other hand, urges us to focus on new innovations and the forces responsible for their creation and diffusion. It anticipates that material substitution will be most pervasive in those end uses experiencing rapid technological change even if the prices of competing materials remain unchanged. So the two views lead to quite different research agendas for understanding the nature and causes of material substitution.

Historical Trends

Demand for iron and steel, aluminum, copper, nickel, lead, zinc, and other major metallic and non-metallic commodities displays two particularly striking characteristics over the past century. First, as Figure 2.9 shows, growth in global production—a good proxy for consumption and demand—has been extraordinary, driven by the surge in world population and the growth in per capita income in many countries over this period. In the case of steel, aluminum, and nickel, new production technologies introduced during the latter half of the nineteenth century and the early years of the twentieth century also spurred demand growth by greatly lowering production costs. As a result, the consumption for each of the major metals over the past century—indeed over the past 50 years—surpasses by a sizable margin their total consumption over all previous history. One of the pressing questions raised by this rapid growth is: will it continue in the future? We will return to this question in Chapter 9 when we address concerns over the long-run availability of non-renewable mineral commodities and examine exploration and the discovery of mineral resources.

Second, since the Industrial Revolution, Europe, North America, and more recently Japan have accounted for the lion's share of the world's demand for the major metals and non-metallics. As recently as 1990, as Table 2.1 shows for copper, these three regions consumed most of the world's output. Over the past two decades, however, this situation has changed dramatically, thanks first to rising demand in Korea, Taiwan, and other developing countries, and more recently to China's rapid economic growth. Some of China's consumption is re-exported in air conditioners and other material-intensive products, which it manufactures and then exports around the world, but much of it goes into the country's expanding infrastructure and domestic consumption. India and other developing countries, many believe, will in coming years accelerate this tendency for demand to shift from developed to developing countries.

The picture for minor metals and some non-metallics including potash is somewhat different. They are often consumed in one or a few applications, and as a result their demand can rise or fall sharply with developments in a single important end use. For example, molybdenum, which the defense industry uses to produce hardened steels, experienced a surge in demand followed by collapse during and after World Wars I and II. During the second half of the twentieth century, this metal found new uses in other industries, reducing its dependence on the defense industry. Similarly, titanium is largely used to produce military and civilian aircraft and as a result watches its fortune rise and fall with aircraft production.

Nevertheless, the demand for most minor metals and non-metallics has also expanded greatly over the past 50–100 years. Indeed, many have become commercial products only during this period. While the shift in demand from the developed to developing countries is taking place for some, others are used in high-technology applications whose production is still largely concentrated in the developed countries.

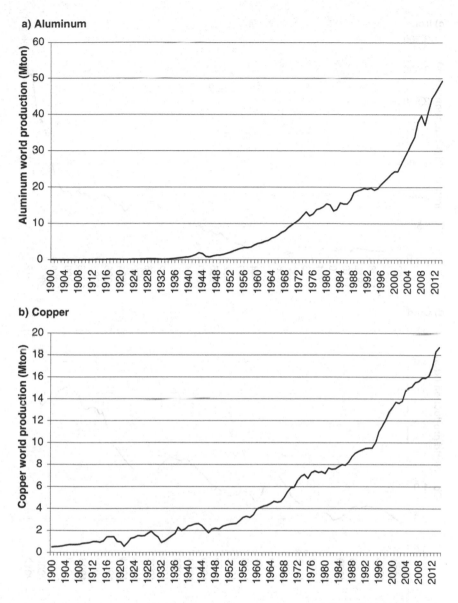

Figure 2.9 World production of aluminum, copper, iron, gold, nickel, and zinc, 1900–2014 (source: US Geological Survey (2014)).

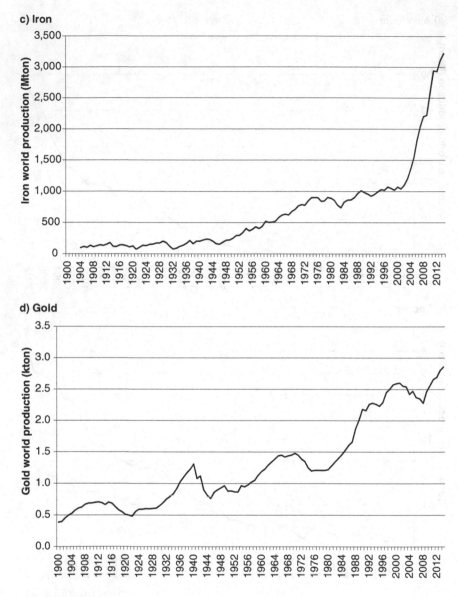

Figure 2.9 continued

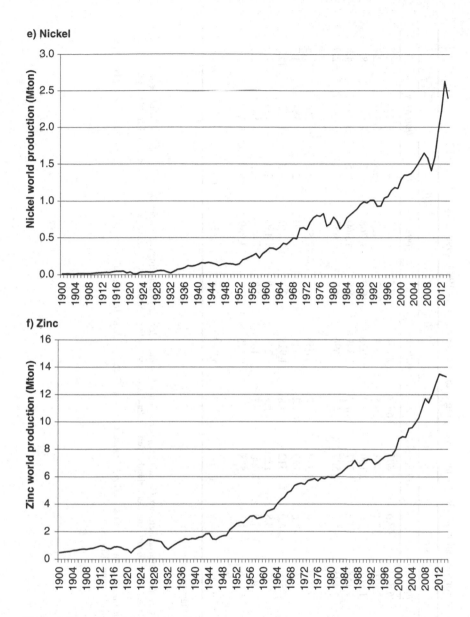

e) Nickel

f) Zinc

Figure 2.9 continued

Table 2.1 Copper consumption for the major developed and developing countries for the years 1980, 1990, 2000, 2010, and 2014

	1980		1990		2000		2010		2014	
	'000 tons	%	'000 tons	%	'000 tons	%	'000 tons	%	'000 tons	%
*Developed countries**	6,335	67.6	7,407	68.8	9,134	60.1	6,608	34.5	6,486	28.5
Western Europe**	2,816	30.1	3,173	29.5	4,063	26.7	3,128	16.3	2,891	12.7
United States	1,868	19.9	2,150	20.0	3,026	19.9	1,768	9.2	1,841	8.1
Japan	1,158	12.4	1,577	14.7	1,349	8.9	1,060	5.5	1,085	4.8
Others	493	5.3	507	4.7	696	4.6	652	3.4	669	2.9
Developing countries	3,029	32.4	3,354	31.2	6,058	39.9	12,526	65.5	16,290	71.5
Brazil	246	2.6	129	1.2	331	2.2	470	2.5	384	1.7
China	370	4.0	512	4.8	1,928	12.7	7,419	38.8	11,352	49.8
India	77	0.8	135	1.3	240	1.6	430	2.2	434	1.9
Russia	1,300	13.9	1,000	9.3	183	1.2	421	2.2	568	2.5
Others	1,036	11.1	1,578	14.7	3,375	22.2	3,789	19.8	3,352	15.6
Total	9,364	100.0	10,761	100.0	15,192	100.0	19,135	100.0	22,776	100.0

Source: World Bureau of Metal Statistics (annual).

Notes:
* The developed countries are Australia, Austria, Belgium, Canada, Denmark, Finland, France, Germany (including the former Democratic Republic of Germany), Greece, Iceland, Ireland, Italy, Japan, Luxembourg, the Netherlands, New Zealand, Norway, Portugal, Spain, Sweden, Switzerland, Turkey, the United Kingdom, and the United States.
** Western Europe includes Austria, Belgium, Denmark, Finland, France, Germany, Greece, Iceland, Ireland, Italy, Luxembourg, the Netherlands, Norway, Portugal, Spain, Sweden, Switzerland, and the United Kingdom.

Highlights

While the nature of demand can vary greatly from one mineral commodity to another, this chapter describes the conventional wisdom that mineral economists believe generally characterizes mineral commodity demand. These are the presumptions with which we normally start inquiries of demand.

In the short run, for example, income or GDP is often the dominant force shaping demand. Fluctuations in the economy can explain much of the year-to-year changes in the demand for steel, zinc, and other mineral commodities. Over the longer run other determinants play a greater role, reducing the dominance of income. So analysts trying to explain the changes in the demand for copper, nickel, and other commodities over the past 25 years almost always focus on changes in own price and the prices of substitutes in addition to trends in income. Other determinants, such as changes in technology, consumer preferences, and government policies, are also often important over the long run. However, they tend to receive less attention, in part because their influence is more difficult to quantify and assess.

Given the importance of income, particularly in the short run but also in the long run, mineral economists in analyzing demand often begin by separating the influence of this variable from that of changes in the intensity-of-use. Intensity-of-use—the ratio of a commodity's consumption over GDP—in turn depends on the product composition of income (the mix of goods and services the economy produces) and the material composition of products (the mix of materials used in the production of individual goods).

The intensity-of-use hypothesis contends that the product composition of income and hence intensity-of-use changes with economic development, first rising with per capita income (or other indicators of development) and then at some point declining. Since material prices, new technology, and the other forces causing the material composition of products to change are less a function of economic development and more a function of time, the inverted U-shaped curve between the intensity-of-use and per capita income, which the intensity-of-use hypothesis predicts, is not stable but shifts over time, normally downward in response to new resource-saving technology.

While this reduces the usefulness of the intensity-of-use hypothesis for forecasting demand, the hypothesis does provide the rationale for believing that the long-run elasticities of mineral commodity demand with respect to income are greater than 1 in developing countries still operating on the upward-sloping segment of the intensity-of-use curve, and less than 1 in developed countries. As a result, a 1 percent increase in GDP in China or India will stimulate the demand for mineral commodities more, possibly much more, than a comparable increase in GDP in the United States or Japan.

In the short run, the income elasticities of demand for most commodities are presumed to be greater than 1 in developed as well as developing countries. This is because their consumption for the most part is concentrated in the construction, transportation, capital equipment, and consumer durable sectors, whose output varies with the business cycle but with fluctuations of much greater amplitude than GDP itself. As a result, the demand curve for most mineral commodities shifts considerably over the business cycle. As Chapter 4 shows, this helps explain the volatility of mineral commodity prices and production over the short run.

Focusing on the role of material prices, the elasticity of demand with respect to own price (as well as the prices of substitutes and complements) is normally presumed to be greater in the long run than the short run. This is because it often takes time, several years or more, for consumers to alter their production lines and carry out the substitution of one material for another. Moreover, some of the impact of price changes on long-run demand occurs indirectly, as a result of new innovations and technologies induced by changes in material prices. As a result, price elasticities of demand normally are presumed to be less than 1 in the short run and greater than 1 in the long run. This means that the demand curves for most mineral commodities in the short run will have steep slopes, much more so than their long-run curves.

Notes

1   Lump ore, typically about 65 percent iron, is extracted from the earth, shipped, and used directly in blast furnaces to produce iron. Ore running 30–35 percent iron is often crushed, upgraded, and then rolled and baked into small, round shapes or pellets at or near the mine. Sinter is produced by heating and fusing iron ore fines into larger pieces, usually at or near the consuming blast furnaces.
2   An interesting exception is potash. Producers sell muriate of potash or MOP in fertilizer bags, which farmers and other consumers use directly.
3   Another more technical reason for using the linear structure found in Equations 2.2 and 2.3 is that under very general assumptions it captures the first-order effects of any function regardless of its complexity. This is known as the Taylor approximation.
4   Still, it is worth noting there are no standard definitions of the short and long runs. Some copper-producing companies, for example, consider five years or more to be the long run. For others the break is at eight or ten years.
5   Estimates of the short-run income elasticity of demand for mineral commodities from econometric and other statistical studies are often less than 1. However, there is evidence to suggest that these results reflect the failure of these efforts to take account of the influence of changes in consumer preferences and new technology on demand. As a result, the negative influence of these variables on demand is captured by the income variable, tending to reduce the estimated effect of an increase in income on metal demand. See Pei and Tilton (1999).
6   Among the first to produce such forecasts were the International Iron and Steel Institute (1972) and Malenbaum (1973, 1978).

References

Demler, F.R., 1980. The nature of tin substitution in the beverage container industries. Unpublished PhD dissertation, Pennsylvania State University, University Park, PA.
Demler, F.R., 1983. Beverage containers, in Tilton, J.E., ed., *Material Substitution: Lessons from Tin-Using Industries*, Resources for the Future, Washington, DC, 15–35.
International Iron and Steel Institute, 1972. *Projections 85: World Steel Demand*, International Iron and Steel Institute, Brussels.

Malenbaum, W., 1973. *Material Requirements in the United States and Abroad in the Year 2000*. A research report prepared for the National Commission on Materials Policy, University of Pennsylvania, Philadelphia.

Malenbaum, W., 1978. *World Demand for Raw Materials in 1985 and 2000*, McGraw-Hill, New York.

Pei, F. and Tilton, J.E., 1999. Consumer preferences, technological change, and the short-run income elasticity of metal demand. *Resources Policy*, 25 (2), 87–109.

Radetzki, M. and Tilton, J.E., 1990. Conceptual and methodological issues, in Tilton, J.E., ed., *World Metal Demand: Trends and Prospects*, Resources for the Future, Washington, DC, 13–34.

Tilton, J.E., ed., 1983. *Material Substitution: Lessons from Tin-Using Industries*, Resources for the Future, Washington, DC.

Tilton, J.E., 1991. Material substitution: The role of new technology, in Nakicenovic, N., Grubler, A., eds., *Diffusion of Technologies and Social Behavior*, Springer-Verlag for the International Institute for Applied Systems Analysis, Berlin, 383–406.

US Bureau of Economic Analysis, 2014. *Current-Dollar and "Real" Gross Domestic Product*. Available at: www.bea.gov/national/index.htm#gdp

US Census Bureau, 2014. *Population Estimates*. Available at: www.census.gov/popest/data/historical/index.html

US Geological Survey, annual. *Mineral Commodity Summaries*. Available at: http://minerals.usgs.gov/minerals/pubs/commodity/zinc

US Geological Survey, 2014. Volume 1, Table 1, in *Minerals Yearbook*. Available at: http://minerals.usgs.gov/minerals/pubs/commodity/myb/#C

World Bureau of Metal Statistics, annual. *World Metal Statistics Yearbook*, World Bureau of Metal Statistics, Ware, UK.

3 Supply

The supply of a mineral commodity reflects the quantities that sellers in the relevant market are willing and able to offer at various prices over a particular period of time, such as a year. Normally, we expect supply to increase with price largely because firms tend to find it profitable to produce more as price rises. As Chapter 2 notes, the relevant market may be global or regional, even local.

Sources of Supply

The nature of mineral commodity supply, like the nature of demand, varies greatly. At the most basic level, as Figure 3.1 shows, the supply of mineral commodities comes primarily from two sources. The first is *primary production*, which entails the extraction of minerals from the earth by mining or other means and their first use by humans. The second is *secondary production*, which involves the collection and recycling of materials that have already been used at least once in the production process. In addition, the depletion of stocks and inventories may for a time add to supply.

Primary production, in turn, can be further broken down. Some mineral commodities, such as iron, phosphate rock, and aluminum (produced from bauxite), are normally extracted as single or *individual products*. Others are

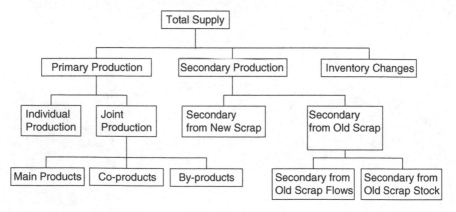

Figure 3.1 Sources of supply for mineral commodities.

produced as *joint products*. Lead and zinc, for example, are often extracted together from the same mine. Molybdenum and gold are often found in porphyry copper deposits. Nickel sulfide mines in Canada and elsewhere may produce copper as well. Rare earth deposits often possess a host of valuable minerals. Potash and lithium are extracted from the same brines around the globe.

Where joint production occurs, main products, co-products, and by-products may be recovered. By definition, a *main product* is so important to the economic viability of a mine that its price alone determines the mine's output of ore. When the prices of two or more joint products affect the level of ore production, they are *co-products*. A *by-product* on the other hand is so unimportant its price has no influence on the ore output of the mine. A by-product must be extracted with either a main product or with co-products.

In the case of secondary production, we normally distinguish between supply produced from the recycling of new and old scrap. *New scrap* is the material recovered in the process of producing mineral commodities and the goods in which they are used. When the tops for beverage cans are punched out of a sheet of aluminum alloy, the skeleton left after the tops are removed is new scrap and can easily be recycled. *Old scrap*, on the other hand, arises when material-containing products reach the end of their useful lives. Once opened and its contents consumed, an empty aluminum soft drink can becomes old scrap.

In assessing the supply of secondary production from old scrap, for reasons we will explore, it is useful to distinguish between *old scrap stocks* and *old scrap flows*. Old scrap flows reflect the amount of old scrap that becomes available for recycling during a given year (or over another time period). Old scrap stocks, on the other hand, contain all of the scrap available for recycling at the start of the year. These stocks contain the material in old scrap flows from earlier years that has yet to be recycled. For example, the amount of old steel scrap available for recycling this year depends on the stock of steel scrap available at the beginning of the year plus the flow of old steel scrap that becomes available over the year from the scrapping of old automobiles and other steel-containing products.

Producing metals and other mineral commodities by recycling scrap is called secondary production, not because recycled or secondary materials are of poorer quality or inferior, but because the scrap from which they are made is not the original or primary source of supply. In contrast, primary production entails the extraction of mineral resources from the earth and their use for the first time.

As noted above, iron ore and bauxite are largely mined as individual products. However, the iron, steel, and aluminum produced from these resources are recycled in substantial quantities. As a result, secondary production is an important part of the total supply for these metals. Copper and nickel are often produced as main products, and again secondary production adds significantly to their supply. Lead and zinc are often co-products. While secondary production is of little import for zinc, it accounts for over half of the supply of lead, due to the widespread recycling of lead in automobile batteries. Gold and silver are often by-products of copper, nickel, lead, and zinc production, though both are main products in some mines. Gallium is largely produced as a by-product in some bauxite mines. Indium is extracted almost exclusively as a by-product of zinc production, while lithium is recovered from certain brines where it is a by-product and potash is the main product.

This chapter looks next at the primary production of mineral commodities— first those extracted as individual products and then those produced as joint products. It then focuses on supply from recycling and secondary production, the derivation of total supply from its component sources, and finally the use of production costs to estimate actual supply curves. We return to the role of inventories and stocks as a source of supply in Chapter 4.

Mining and Primary Production

Primary production requires four quite distinct steps or activities. The first entails exploration and the discovery of identified resources. Identified resources are mineral deposits that may or may not be economic to exploit. The second step involves the transformation of identified resources to reserves. By definition reserves encompass the quantity of a mineral commodity contained in deposits that are both known (that is, discovered) and economic to exploit given current technology, the price of the commodity, and other prevailing conditions. We will examine the nature of mineral resources and reserves more fully in Chapter 9.

At this point, it suffices to note that reserves can be created in three ways: (1) by the discovery of new economic deposits, (2) by new technology, better management, lower input prices, or other developments that reduce costs, making previously known but uneconomic deposits now profitable to exploit, and (3) by higher commodity prices that similarly makes previously known but uneconomic deposits profitable to exploit. Exploration that leads directly to the discovery of reserves simultaneously covers the first and second steps.

After the creation of mineral reserves, the third step in the process entails investing in production capacity—the development of new mines and the construction of mills, smelters, refiners, chemical plants, railroads, harbors, ships, and the other facilities needed to take the ore from the ground and to transform it into a refined metal or otherwise marketable mineral commodity.

The fourth step is the actual production of mineral commodities, such as refined copper or cold rolled steel sheet. Because production capacity is fixed in the short run—determined by the investments in mines and processing facilities carried out during earlier periods—this fourth step depends on how much capacity producers possess and how much of that capacity they elect to use. For this reason, the supply of primary production this year depends on current and past market conditions. In addition, current market conditions, because they affect today's efforts to create new reserves and new production capacity, influence future as well as current supply. While particularly attractive new discoveries may be developed and exploited within a relatively short period—five years, for example—history also suggests that many discoveries do not contribute to supply for decades (if ever).

Mineral economists frequently note that the creation of identified resources and reserves—the first two steps in the production process—are unique to mineral commodities. The production of other goods typically involves just two steps—the creation of capacity followed by the production of the good itself. Manufacturing automobiles, for example, does not entail production steps similar to the creation of identified resources and reserves.

Just how significant this difference is, however, is unclear. New technology, we have noted, often creates new reserves by making it possible to exploit previously known but uneconomic mineral deposits. In a similar way, one might argue that the first step in the production process in many high-tech industries is the use of science and technology to innovate and develop new and better semiconductors, computers, mobile phones and other such devices. Similarly, a rise in lumber prices may convert a standing forest whose timber was uneconomic to cut from an identified resource to reserves. A stronger case perhaps can be made that it is just the first step—exploration and the discovery of identified resources—that distinguishes the production of mineral commodities.

The field of mineral economics has explored in some depth the supply of what we might consider the intermediate products in this four-step production process—identified resources, mineral reserves, and new capacity. This chapter, however, focuses largely on the supply of mineral commodities at the end of this four-step production process.

Individual Products

In analyzing the supply of individual mineral commodities, we look at the determinants of supply, the supply function and supply curve, and the elasticity of supply.

Determinants of Supply

Just which variables do we need to consider in order to understand why the supply for mineral commodities varies from one year to the next and over the longer run? The answer depends on the commodity, the time horizon of the analysis, and other factors, and calls for considerable judgment on the part of the analyst. Though no single list is appropriate for all situations, the following are often important, particularly over the long run.

Own price. Competitive firms have an incentive to increase their output up to the point where the costs of producing an additional unit just equals the extra revenue they receive from selling that unit. As a result, a rise in the price should normally increase supply, and a fall in price reduce it.

In the short run, we know that existing capacity may constrain the response of supply to a rise in price. It takes time to develop new mines and build processing facilities. Often, producers need five years or longer to respond fully to a price increase. Even more time may be needed to respond fully to a price decrease. Mining and mineral processing are capital-intensive activities, requiring equipment and facilities that last for decades. So firms will remain in production, even when price falls below average total costs, as long as they are recovering their variable or out-of-pocket costs. Only when existing capacity needs to be replaced will they shut down.

Input costs. The costs of labor and other inputs used in mining and processing also affect profitability, and in turn supply. Japan, for example, during the 1960s built one of the world's largest aluminum smelting industries. Aluminum smelting uses huge amounts of electricity, and Japan at that time produced much of its electricity from imported oil. When oil prices jumped sharply in the 1970s, it quickly became clear that smelting in Japan no longer made economic sense, and the country proceeded to largely dismantle its aluminum industry within a decade. Aluminum smelting in Japan is a rather dramatic example of the impact that changing input costs can have on supply, and is unusual in terms of the speed with which supply responded. More typically, the full effect of changing input costs on supply takes years or even decades as producers are often willing to continue production until existing capacity becomes obsolete or for other reasons needs to be replaced.

Technological change. Advances in technology that reduce mining and processing costs also can affect mineral supply. During the 1980s and 1990s, for example, the development and diffusion of solvent extraction electrowinning (SX-EW) in the copper industry, first in the United States and then in Chile (and to a lesser extent elsewhere), allowed the recovery of copper from previously uneconomic oxide minerals.

Strikes and other disruptions. During the 2003–2013 boom, strikes by workers in Chile and other copper-producing countries, who felt they were not sharing sufficiently in the wealth created by significantly higher copper prices, hampered company efforts to increase output. Pushing existing capacity to its limits can also cause accidents and other problems, as the slope failure that interrupted copper production at the Bingham Canyon mine in the United States in 2013 illustrates. In the aluminum industry, which depends heavily on hydroelectric power, supply has often been curtailed by inadequate rainfall. In Africa and elsewhere, supply has suffered from political unrest and civil war. Strikes, mine accidents, natural disasters, civil disturbances, and other such disruptions affect supply of mineral commodities not only by reducing production but also by interrupting transportation and shipment.

Government activities. Public regulations and tax policies can substantially raise production costs, and if changed frequently, increase the perceived risk for mining companies. Some countries have required mining companies to provide the government with equity shares, to build public infrastructure, to favor domestic suppliers, to carry out downstream processing within the country, and to employ nationals. Such requirements may reduce efficiency and increase the costs of mining and mineral processing. Zimbabwe in recent years provides a particularly distressing example of how government policies and activities can adversely affect mineral supply.

On the other hand, governments can also stimulate supply. China, for example, has historically encouraged the domestic production of rare earths and other minerals with lax environmental regulations and other measures that subsidize new mines and processing facilities. The United States, China, and other countries maintain strategic stockpiles. When they reduce their inventories, they add to the available supply.

In addition, between 1960 and 1990, many governments nationalized their mining operations and created state mining companies. After 1990 there was a reversal in this trend in the metal industries (though not in the oil industry). Still, some state mining companies continue to exist, including Codelco (Corporación Nacional del Cobre), which is owned by the Chilean government and is the world's largest copper producer. In their production and investment decisions state mining companies may be less concerned about profits than maintaining employment and other public goals. If so, their supply presumably responds less to price signals, particularly to low prices during a downturn in the market.

Other government actions over the past several decades that markedly altered the nature of supply for many mineral commodities include the decisions by China, Russia, and other member states of the former Soviet Union to move from centrally planned to market economies. In the process they removed barriers that previously isolated their economies from world trade and the global economy. The result has been a tremendous geographic expansion for many mineral commodity markets, which among other things has enhanced competition by increasing the number of producers and consumers.

Market structure. Many mineral commodity markets are competitive in the sense that no producer or supplier possesses the ability and the will to control the market price. In such markets, producers have an incentive to increase the supply they offer to the market as long as the price they receive for one more unit exceeds their costs. Where just a few firms (or countries) account for most of a commodity's production, the market may not be competitive. In such cases, the major firms may maintain a producer price (that is, a price at which they are willing to sell all of their output and below which they are willing to sell nothing) or possibly a cartel price. We examine in greater detail the importance of market structure in Chapter 7.

The Supply Function

The relationship between the supply of a mineral commodity and its principal determinants, such as those just discussed, is given by the *supply function*. Normally, it is expressed mathematically. Equation 3.1, for example, is the function for a commodity whose supply during a given year t (Q_t^s) depends on its price (P_t^o), the wage rate paid by producers (W_t), the cost of energy (E_t), and strikes (S_t).

$$Q_t^s = g(P_t^o, W_t, E_t, S_t) \tag{3.1}$$

This is a rather simple supply function. It does not consider technological change and other variables that often affect supply. It contains no lagged values for price or the cost variables, and hence takes account only of their short-run influence. Finally, its exact specification is not indicated.

The Supply Curve

The relation between a mineral product's price and its supply is often of special interest. It is portrayed by the *supply curve*, which shows how much producers and others will offer to the marketplace at various prices over a given time period, such as a year, on the assumption that all the other variables affecting supply remain at some specified level.

The supply curve is normally drawn sloping upward, indicating that supply increases with price. In special circumstances, however, the curve can be horizontal, vertical, or even downward sloping. A horizontal supply curve implies that sellers are willing to provide the market with as much as they have to offer at a particular price and with nothing below that price. We have noted above that this occurs when firms maintain a producer price. A vertical supply curve implies that sellers will provide the market with a given quantity, no more and no less, regardless of the price. This may occur when firms are operating at full capacity and the price increases. A downward-sloping supply curve means that sellers will offer a greater quantity to the market at a lower price. This might occur if state mining companies feel obliged to increase output when price falls in order to help maintain their countries' foreign exchange earnings.

While such situations do occur, one normally expects the supply curve to be upward sloping, and in this important respect different from the demand curve. Many other characteristics of the supply curve, however, are the same or similar to those discussed in Chapter 2 for the demand curve.

For example, a movement along the supply curve reflects a change in price, while a shift in the curve itself reflects a change in any of the other determinants of supply. Similarly, supply, like demand, may differ from production and consumption due to changes in stocks and inventories.

There are also many different supply curves. Again, like demand curves we can divide them into two groups: the market supply curve, which indicates the quantity of a mineral commodity that all sellers are willing to offer to the market over a year or some other period of time; and all other supply curves. The latter in the case of refined copper, for example, include the curves for the supply of all Chilean copper producers, for the supply of Codelco (the state mining company of Chile), for the supply of El Teniente (one of Codelco's mines), for the supply of Chinese copper imports, for the supply from US government stockpiles, and so on.

Ultimately, we are usually interested in the market supply curve. Other supply curves are of interest largely because of the insights they provide regarding the market supply curve. Where the market supply and demand curves intersect, the quantities provided by sellers and the quantities desired by buyers are equal, and at that price the market clears.

The supply curve assumes that producers have a certain amount of time to adjust to changes in price, input costs, and other determinants. Here, as with demand, we usually distinguish between the short and long run. In the *short run*, firms can alter their production but not their capacity. In the *long run*, they can change both. There is sufficient time to construct new mines and processing facilities by discovering and developing previously unknown deposits and by developing previously known but uneconomic deposits that become profitable thanks to higher prices or new technologies. In shrinking rather that growing markets, there is also time over the long run for firms to reduce capacity by closing down facilities as they require new investments to remain in operation.

In order to assess the general shape of supply curves in the short and long run, we need to distinguish two types of commodity markets—producer markets and competitive markets. Firms in producer markets set a price at which they are prepared to sell their output. These markets, which tend to have just a few major sellers, have relatively stable prices, though when demand is weak, actual prices may fall below the set or quoted price as a result of discounting and other concessions. Many steel products, for example, are sold in producer markets, as are potash, lithium, iodine, precious gemstones, and many minor metals.

In competitive markets, prices are determined by the interplay of supply and demand, and fluctuate to ensure that the market clears. Many buyers and sellers participate in competitive markets, and price is often set on a commodity exchange, such as the London Metal Exchange (LME). Silver, gold, tungsten, lead, and zinc, in addition to aluminum, copper, and nickel, are examples of metals currently traded on competitive markets. In competitive markets producers are price takers and have little influence over the going market price. They do, of course, still control their own supply.

Figure 3.2 shows short-run market supply curves for sellers in a competitive market and in a producer market. The curve for the competitive market indicates that no supply is forthcoming at very low prices (below price P_0) as sellers withhold supplies in anticipation of higher prices in the future. In addition, firms have an incentive to stop production when price falls below their average variable or out-of-pocket costs. At some point, as the price rises, supply begins to enter the market. Once this threshold is reached, supply at first expands greatly in response to higher prices as production increases and more of what is produced is offered for sale. Eventually, however, further increases in supply must come from producer inventories. Since producers will deplete their stocks only at high prices, and since stocks can normally augment supply by only modest amounts compared to production, the supply curve for the competitive market shown in Figure 3.2 becomes quite steep at high prices and finally vertical. As output approaches the capacity constraint, firms are unwilling or unable to add further to supply.

Figure 3.2 also shows a short-run supply curve for a commodity sold in a producer market. It is simply a horizontal line at the producer price (P^*) that extends from zero to the quantity that existing capacity can produce (plus any possible reductions in the stocks held by producers and others). Since firms cannot increase their capacity in the short run, current capacity imposes a constraint on supply. Figure 3.2 shows the supply curve for a producer market stopping at this barrier.

The curve shown in Figure 3.2 assumes that firms faithfully adhere to the producer price. If this is not the case, if some or all firms offer discounts off the producer price during a period of weak demand, the curve is not perfectly horizontal, but instead drops somewhat at lower quantities, reflecting these price concessions.

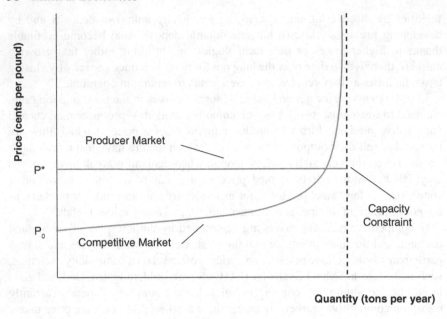

Figure 3.2 Supply curves in the short run.

Figure 3.2 shows that the short-run supply curve for a competitive market lies mostly below the short-run curve for a producer market. This is because in a competitive market producers are willing to provide their output to the market as long as they are covering their variable costs or cash costs. In producer markets, firms generally take into account their fixed as well as variable costs in setting the producer price. As Figure 3.2 shows, however, as output approaches the capacity constraint, the curve for the competitive market rises above the curve for the producer market. This is because the variable or cash costs of additional production rise sharply as output approaches the capacity constraint. While firms in competitive markets raise their prices with this rise in costs, firms in producer markets do not.

In the long run, firms can expand the capacity of existing operations and construct new mines and processing facilities. They have time to conduct exploration and generate new reserves. They can create new technologies that permit the exploitation of new types of deposits. As a result, there is no capacity constraint in the long run.

It is true that the earth is finite, and so the amounts of tin, platinum, iron, and other mineral products it contains are finite. However, as we will see in Chapter 9, the quantity of every mineral commodity found in the earth's crust is so enormous compared to the amount supplied annually (which is what the supply curve typically relates to price) that this ultimate constraint is not relevant.

As a result, as Figure 3.3 shows, the long-run supply curve for a competitive market is likely at first to rise significantly, reflecting the limited number of low cost, world-class deposits. As price continues to increase, however, it does not at some point encounter a capacity constraint forcing it to turn upward and become vertical. Instead, the long-run supply curve for many commodities becomes flatter and hence more elastic at higher prices and larger quantities. This is because more costly sources of supply or deposit types are found in greater numbers and contain greater quantities of the resource. Large porphyry copper deposits containing 0.4 percent copper equivalent, for example, are fairly abundant.[1] Many have been discovered over the past 50 years, even though exploration efforts have not focused on the discovery of such marginal deposits. At some price for copper, however, such deposits will become attractive exploration targets and presumably many more will be discovered, allowing supply to expand greatly.

Figure 3.3 also portrays a long-run supply curve for a producer market. Again, it is a horizontal line at the producer price (P*). However, it does not terminate at a capacity constraint like the short-run curve but rather continues indefinitely to the right. Moreover, producer prices tend to change from time to time in response to changes in production costs and other considerations. So the long-run supply curve for a producer market is likely to shift up and down over time.

Figure 3.3 shows the long-run supply curve for the producer market above the curve for the competitive market. This suggests that producer prices will normally be higher than competitive prices. While this may be the case, there is still much we have to learn about how producer prices are actually set. While average total costs may often play an important role, other factors are presumably important as well. As a result, we should consider the relative heights of the two long-run supply curves shown in Figure 3.3 simply as plausible and illustrative.

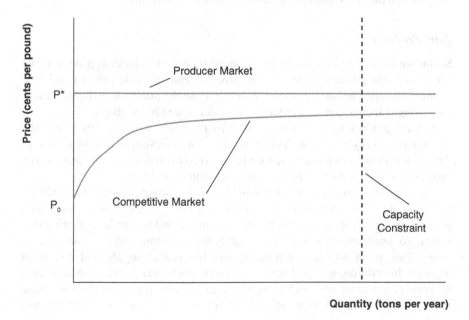

Figure 3.3 Supply curves in the long run.

The Price Elasticity of Supply

To measure the responsiveness of supply to changes in price, mineral economists use the price elasticity of supply, defined as the partial derivative of supply with respect to price times the ratio of price to supply, as shown in Equation 3.2a. This measure, as Equation 3.2b indicates, reflects the percentage change in supply caused by a 1 percent change in price.

$$E_{Q_t^s, P_t^0} = \frac{\partial Q_t^s}{\partial P_t^0} \cdot \frac{P_t^0}{Q_t^s} \tag{3.2a}$$

$$= \frac{\text{percent change in } Q_t^s}{\text{percent change in } P_t^0} \tag{3.2b}$$

We say that supply is elastic when the elasticity is greater than 1, and inelastic when it is less than 1. Where the supply curve is vertical or where it simply ends (as in the case of producer markets in the short run), supply is completely unresponsive to price and the supply elasticity is 0. Where the supply curve is relatively flat or horizontal, supply is highly or infinitely responsive to price and the elasticity is very large.

Normally, mineral economists and industry analysts assume that the price elasticity of supply is greater in the long run than in the short run, as producers cannot add to capacity in the short run. This is true, however, only when output in the short run is approaching capacity. When capacity utilization is low, supply can be quite responsive to changes in price, even in the short run.

Joint Products

So far we have focused only on the primary supply of individual products. We now turn to the primary supply of joint products and ore bodies from which two or more mineral commodities are recovered. In such situations, as noted earlier, we distinguish among main products, co-products, and by-products.

A main product by definition is so important that it alone determines the economic viability of a mine. As a result, main product supply is quite similar to that for individual products, which we have already considered. This section as a result will focus on the supply of by-products and co-products.

Before proceeding, however, we should note that some mineral commodities, such as gold, silver, molybdenum, iodine, and nitrates, are main or individual products at some mines, co-products at other mines, and by-products at still other mines. To determine the total primary supply for these commodities, one needs to assess their main product, co-product, and by-product supply and add them together. In addition, gold and other mineral products may at the same mine be a by-product at a given time and a co-product or even main product at other times, if the prices of gold and the associated joint products vary sufficiently over time.

By-product Supply

A by-product is produced with a main product or with co-products. It may also be extracted in association with other by-products. By-product supply differs in two important respects from the supply of individual or main products.

The first is that by-product supply is limited by the output of the associated main product (or co-products). The amount of molybdenum recoverable as a by-product of copper production, for example, cannot exceed the physical quantity of the molybdenum actually in the extracted copper ore. As production approaches this constraint, the by-product supply curve turns upward and becomes vertical. This is because a higher by-product price does not increase the output of the extracted ore or of the main product. Otherwise, it would not be a by-product. Thus, at some output, the supply of a by-product even in the long run becomes unresponsive or inelastic to further increases in its price.

Figure 3.4 illustrates this characteristic of by-product supply. The long-run supply curve is quite elastic until output approaches the by-product constraint determined by main product output. Thereafter, little or no increase in supply is possible. Since normally by-product producers are competitive firms, Figure 3.4 shows the by-product supply curves for only the competitive market. The same constraint, however, would exist in a producer market.

The by-product constraint, it is important to note, shifts in response to changes in the demand and supply of the main product. If the demand for copper goes up, for example, this causes the price and output of the main product, copper, to rise, increasing in the process the amount of extracted ore from which by-product molybdenum can be recovered. For this reason, the supply function for a

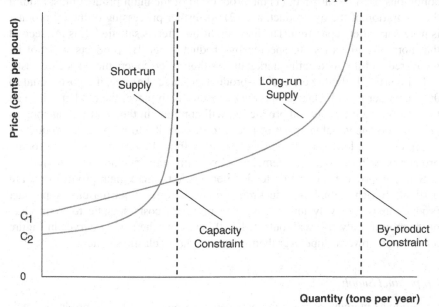

Figure 3.4 Supply curves for a competitive by-product producer.

by-product should normally contain the output (or price) of the main product, in addition to its own price and the other important supply variables discussed above in connection with the supply of individual products. While changes in the by-product's own price reflect a movement along its supply curve, a change in the price or output of the main product produces a shift in both the by-product constraint and the by-product supply curve.

Main product ores may contain more of a by-product than existing capacity can treat. In such situations, as Figure 3.4 shows, a capacity constraint imposes even more restrictive limits than the by-product constraint on supply in the short run. In the long run, of course, new capacity can be built and the binding constraint becomes the by-product constraint.

The second important difference between by-product and individual (or main) product supply is that only costs specific to by-product production affect by-product supply. Joint costs, those necessary for the production of both the main product and the by-product, are borne by the main product and do not influence by-product supply.

This means that by-product supply curves for competitive markets, such as those shown in Figure 3.4, reflect the marginal costs of by-product production exclusive of all joint costs. As a result, by-product supply is often, though not always, available at lower costs than main or individual product supply, at least until by-product supply approaches the by-product constraint.

It is sometimes said that by-products are basically free goods and that the by-product supply curve is simply a vertical line at that output reflecting the amount of by-product in the main product ore. For this to be true, however, two conditions must be satisfied: (1) the production of the main product must require the separation of the by-product, and (2) no further processing of the by-product is necessary after separation. The first condition often is satisfied. It is the second that normally gives rise to specific by-product costs. By-products will not be recovered and supplied to the market unless their price covers these specific costs.

In Figure 3.4, the lowest cost by-product producer has specific costs equal to $0C_1$ cents per pound. Once by-product capacity is in place, the market price can fall below specific costs and production will continue in the short run, as long as price covers the variable or out-of-pocket costs specific to by-product production ($0C_2$). Over the long run, however, capacity will not be replaced and by-product production will cease if price remains below minimum specific costs ($0C_1$).

Since by-product production tends to occur first where main product ores are particularly rich in the by-product mineral or where for other reasons by-product production is relatively inexpensive, the marginal costs specific to by-product production usually rise with output. For this reason, the supply curves in Figure 3.4 have an upward slope over their relatively flat or elastic segment.

Co-product Supply

Co-products in many respects fall between by-products and main products. Their price influences mine output, but so do the prices of other co-products. Joint production costs must be shared, as normally no single co-product can support them alone. As a result, a co-product's price must cover its specific production costs plus some (though not all) joint costs.

This means that a co-product's supply function includes its own price, the prices of associated co-products, its specific costs, and joint costs. A change in any of these, other than own price, causes a shift in the co-product's supply curve. An increase in specific or joint costs, for example, shifts the curve upward, while a rise in the price of associated co-products shifts it downward.

Co-product supply curves have the same general shape as those shown for individual products in Figures 3.2 and 3.3. Co-product supply is similarly constrained by capacity in the short run but not the long run. Since co-products must bear only a part of the joint costs, supply may be available at lower costs than from main or individual product output.

Recycling and Secondary Production

Secondary production adds to mineral commodity supply by recycling new and old scrap. New scrap, as noted earlier, arises in the manufacturing process for new goods. Old scrap comes from consumer and producer goods that have reached the end of their useful lives because they are worn out, obsolete, or for some other reason no longer of use. In recent years, according to the US Geological Survey (annual), the recycling of old scrap alone has accounted for roughly 35, 10, and 70 percent of US aluminum, copper, and lead consumption, respectively.

Secondary supply differs from primary supply in several important ways. For instance, secondary producers are almost all highly competitive and rarely support a product price. As a result, we need not be concerned about secondary supply curves for a producer market.

In addition, scrap availability is what limits secondary supply in the long run. Since the important determinants of the availability of new and old scrap differ, we need to consider the secondary supply from these two types of scrap separately.

Secondary Supply from New Scrap

The availability of new scrap depends on three factors—the level of a commodity's current consumption, the allocation of this consumption among various end uses, and the percentage of consumption in each end use that ends up as new scrap. For example, a jump in copper consumption due to an increase in GDP or a change in consumer preferences increases the availability of new copper scrap. This causes both the availability-of-new-scrap constraint and the long-run supply curve for secondary copper produced from new scrap to shift to the right. In contrast, improved manufacturing techniques that reduce the amount of copper scrap generated in the manufacturing of electric wire or other fabricated products shifts the constraint and the curve to the left.

This means that the supply function for secondary from new scrap, in addition to technological change and other determinants discussed earlier, should include a variable for the availability of new scrap. Often the consumption of a mineral commodity is used as a proxy for the availability of its new scrap. This is usually appropriate in the short run, when the allocation of consumption among end uses and the proportion of consumption resulting in new scrap for each end use are likely to remain more or less constant. However, when these variables change, as is often the case over the long run, they, too, belong in the supply function.

Figure 3.5 shows a long-run supply curve for secondary supply from new scrap. Its shape reflects the cost of collecting, identifying, and processing new scrap. The scrap that is the least costly to recycle will be used first. These costs, given in Figure 3.5 as $0C_1$ cents per pound, determine the point where the curve intersects the vertical axis. As most new scrap is relatively inexpensive to recycle, the slope of the curve rises from this point very gently over the range of possible outputs. Only as output approaches the availability-of-new-scrap constraint does the supply curve turn sharply upward.

The low cost of recycling new scrap compared to alternative sources of supply means that all (or almost all) new scrap is recycled. So over the range of normal prices, little additional supply from new scrap is possible, making this source of supply price inelastic. However, at very low prices, as Figure 3.5 shows, supply is quite elastic with respect to price.

Because most new scrap is attractive to recycle, the capacity constraint, which limits supply in the short run, seldom differs significantly from the long-run constraint imposed by the availability of new scrap. So while the short-run curve lies below the long-run curve—since in the short run firms will recycle new scrap as long as they are recovering their average variable costs (rather than their

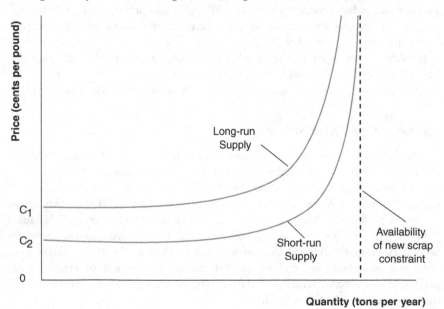

Figure 3.5 Supply curves for secondary production from new scrap.

average total costs)—both curves turn upward and become more or less vertical at about the same output.

Secondary Supply from Old Scrap

The availability of old scrap during any particular year depends on: (a) the flow of material-containing products reaching the end of their useful lives during the year, and (b) the stock of material-containing products no longer in use at the beginning of the year. The number of old automobiles available for recycling during any particular year, for example, includes those scrapped during the year as well as those scrapped in earlier years but which for one reason or another have not yet been recycled.

Old Scrap Flows

The flow of old scrap depends on the number and types of goods in use throughout the economy at the beginning of the year, their material composition, their age distribution, the mean age at which they come to the end of their service life, and the frequency distribution around this mean. Since these variables determine the flow of old scrap, either they or the flow of old scrap can be included in the supply function for secondary supply produced from old scrap flows.

Figure 3.6a shows a long-run supply curve for secondary supply from old scrap flows. At price P_1, this figure indicates that the quantity $0q_1$ of the material in the incoming flow of old scrap is recovered. The rest of the incoming flow is not recycled, but rather added to the stock of old scrap available for recycling in the future. This curve, like secondary supply from new scrap, is constrained by the availability of scrap, though now the constraint is the availability of the material in old scrap flows rather than in new scrap.

The long-run curve begins at the vertical axis at a fairly low price, because some old scrap is of very high quality and can be recycled at a relatively low cost ($0C_1$). However, in contrast with new scrap, costs rise notably as more and more of the old scrap flow is recycled. This is because some scrap is scattered geographically, and so collection costs are high. Some is a mixture of various kinds of scrap, and requires expensive identification and sorting techniques. Some is highly contaminated with rubber, glass, wood, and other waste material, and treatment costs as a result are high. Indeed, in some cases, such as lead in lead-based paints, the material is so dissipated after use that recycling is simply too costly to contemplate. For such reasons, old scrap flows even for gold and platinum are never completely recycled.

Figure 3.6a also illustrates a short-run supply curve for secondary supply produced from the flow of old scrap. It lies beneath the long-run curve at prices below P_1, since secondary producers will continue to operate and supply the market in the short run as long as they can cover their variable costs. Since fixed costs typically account for a small share of total production costs, particularly in comparison with primary production, the short-run curve is drawn close to the long-run curve.

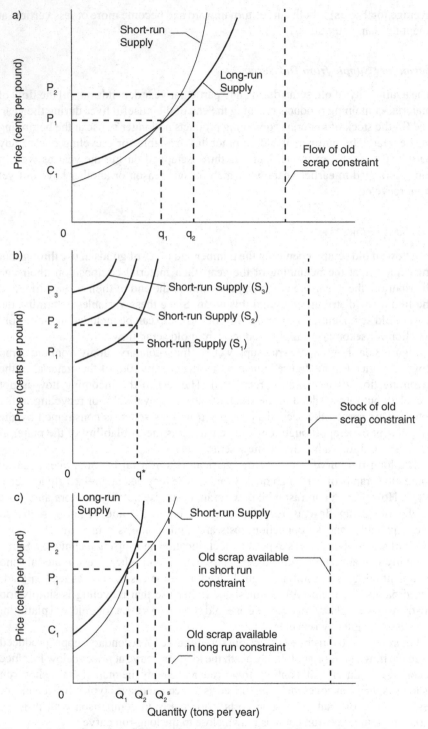

Figure 3.6 Supply curves for secondary production from old scrap: (a) secondary metal from the flow of old scrap; (b) secondary metal from the stock of old scrap; (c) secondary metal from all old scrap.

At prices above P_1, according to Figure 3.6a, the short-run curve is above the long-run curve because at some point producers in the short run encounter a capacity constraint. However, capacity tends to be more fungible in secondary than primary production. It is easier to raise output by increasing the number of shifts and by augmenting labor in other ways. For this reason, the short-run curve above price P_1 is again drawn quite close to the long-run curve.

This, though, need not always be the case. Under certain circumstances, the short- and long-run curves may lie quite far apart. For example, if various end-of-life products that account for a large part of the total old scrap flow would become economic to recycle at prices above P_1, the current market price, and if the capacity to recycle these products does not exist (since they are not at the present time economic to recycle), the long-run supply curve may lie far to the right of the short-run curve at prices above the current market price.

It is also important to note that short-run supply can exceed long-run supply at high prices, with the reverse being true at low prices. This possibility arises as a result of the *accelerated scrapping phenomenon*. This occurs when products close to the end of their service life are scrapped earlier than otherwise as a result of high commodity prices. For example, at times, old or obsolete machines are held in reserve for use during unexpected production booms, for emergencies, or are stored away and cannibalized for their parts. High scrap prices encourage the premature recycling of such equipment. When prices are low, the costs of keeping such equipment in terms of the scrap revenues foregone are quite modest. This means that in some circumstances, the flow-of-old-scrap constraint may not be invariant to price in the short run (and so may not be vertical as shown in Figure 3.6a), but rather may increase with price at least over a range.

Old Scrap Stocks

The stock of old scrap depends on the accumulated past flows of products coming to the end of their useful lives. From this total, the quantities already recovered through recycling must be subtracted. This means that over time the stock of old scrap will grow if the amount recycled is less than the incoming flows.

Even though it depends on two flow variables—the accumulated flow of old scrap and the accumulated recycling of this scrap in the past—the stock of old scrap is, as stated, a stock and not a flow. This has several important implications and accounts for the need to consider secondary supply from old scrap stocks separately from secondary supply from old scrap flows.

Figure 3.6b shows three short-run secondary supply curves from old scrap stocks. The first curve S_1 indicates that at the price P_1 no material is recovered from the stock of old scrap. The explanation for this is that the market price either currently is or recently has been P_1, and all the stock of old scrap profitable to recycle at this price has already been processed. All that is left in the available stock is poorer quality scrap whose recycling costs exceed the price P_1. At higher prices, however, some of the old scrap stock can be economically processed. At P_2, for example, the old scrap stock will contribute the quantity $0q_{1}^{*}$.

This output, however, is available only over the short run. If the price rises from P_1 to P_2, the old scrap stock recoverable at costs at or below P_2 is recycled and so depleted often during the year of the price change. If the price then remains at P_2, the short-run curve changes during the second year in a manner similar to that illustrated by the curve S_2. This new short-run supply curve is constructed by shifting the curve S_1 to the left by the amount recycled ($0q_2^*$), so that it now intercepts the vertical axis at price P_2. To this one must add the new scrap flow occurring during the previous year that was not recycled because it was not economic to do so at the price P_2. Below price P_2, the new supply curve S_2 indicates that no supply from the stock of old scrap is now forthcoming.

Similarly, if the price rises from P_2 to P_3, the short-run supply curve will soon approach the shape indicated by the curve S_3. This means that normally there exists no long-run supply curve for secondary from the stock of old scrap.

The availability of a metal or other material in the stock of old scrap does impose a constraint on the ultimate supply of secondary from old scrap stocks. This constraint, however, is not nearly as binding as in the case of the supply of secondary from new scrap or from old scrap flows. Long before the stock of old scrap is exhausted, recycling is curtailed by high costs. Most of the available old scrap stock has not been recycled precisely because it has been uneconomic to do so. Other sources of supply—individual product, main product, co-product, by-product, secondary from new scrap, and secondary from old scrap flows— have for the most part been cheaper. As a result, secondary supply from old scrap stocks is typically quite modest compared to secondary supply from old scrap flows. Indeed, Gómez et al. (2007) argue that for copper the supply of secondary from old scrap stocks is so modest it can probably be ignored.

Secondary supply from old scrap, both old scrap flows and old scrap stocks, can be derived by combining, or more precisely by adding horizontally, the secondary supply curves for old scrap flows and for old scrap stocks. This is illustrated in Figure 3.6c.

The long-run curve is simply that for secondary supply from old scrap flows, as there is no long-run secondary supply from old scrap stocks for the reasons just explained. The short-run curve is derived by adding the appropriate short-run curve for the stock of old scrap, assumed to be the curve S_1 in Figure 3.6b, to the short-run curve for the flow of old scrap. Since the short-run curve for the stock of old scrap intersects the vertical axis at P_1, implying that below this price no secondary from old scrap stocks is forthcoming, the short-run secondary curve for all old scrap below this price is simply the short-run curve for the flow of old scrap.

Figure 3.6c highlights two interesting facets of secondary supply from old scrap. First, the constraint imposed by the availability of old scrap is actually less binding in the short run than in the long run. This, of course, is because the stock of old scrap, if exploited in the short run, is not available in the long run.

Second, an increase in price, for example from P_1 to P_2, may actually produce an increase in supply that is greater in the short run than in the long run. In the short run, some of the stock of old scrap may be recycled, adding to supply. It is presumably for this reason that Fisher et al. (1972), as well as other studies, have found the price elasticity of secondary supply to be greater in the short run than in the long run, which is just the opposite of what one normally finds with other sources of supply.

However, as Figure 3.6c indicates, this unusual result should be expected only when the market price rises—from P_1 to P_2, for example—so that secondary supply from old scrap stocks is forthcoming. When this is not the case, when the market price declines, the figure suggests that a change in price will produce a greater decrease in supply in the long run than in the short run.

Total Supply

Individual products, main products, co-products, by-products, and secondary from new and old scrap are all potential contributors to a mineral product's total supply. To derive the total supply curve—which is the market supply that encompasses supply from all sellers—we must add horizontally the individual curves for all significant sources (in a manner similar to that just used to derive the secondary supply curve for all old scrap). For example, if the long-run by-product and main product supply curves for a commodity, such as molybdenum, are as shown in Figure 3.7a and if other sources of supply are unimportant and can be ignored, the horizontal summation of these two curves gives the total long-run supply curve, as shown in Figure 3.7b.

Among its many uses, the total supply curve provides an indication of the relative competitiveness of the different sources of supply and, in turn, their relative importance at different market prices. For example, for the commodity whose supply curves are shown in Figure 3.7, by-product production is initially the most competitive and cheapest source of supply. Indeed, when market price is below P_1, it is the only source of supply. As the market price rises above P_1, however, main product production begins and becomes increasingly important. At the price P_2, main product supply exceeds by-product supply.

In deriving the total supply curve, a question at times arises as to whether the secondary supply curve for new scrap should be included. New scrap, some argue, is generated in the manufacturing process and depends on other sources of supply. If fabrication becomes more efficient and generates less new scrap, this does not mean total supply has declined.

Whether to include secondary supply from new scrap depends on the purpose of the analysis. In assessing the extent to which the United States is vulnerable to interruptions in supply from certain foreign producers, for example, we would not want to include secondary supply from new scrap. To do so would ignore the fact that this material is generated from other sources of supply, and so would not be available in their absence. Including it would underestimate US dependence on foreign producers. On the other hand, in assessing the competitiveness of secondary markets, we would normally want to consider secondary supply from new scrap.

Regardless of how secondary supply from new scrap is treated, it is important that total supply and demand be consistent in this regard. If the total demand curve takes account only of material actually contained or embodied in final products,

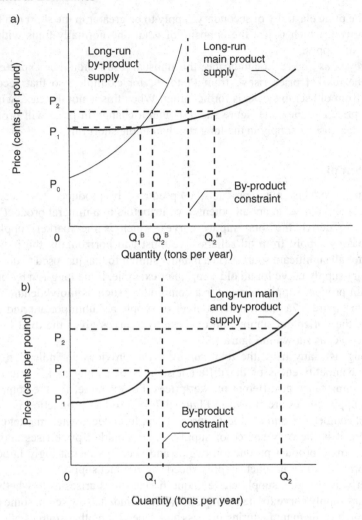

Figure 3.7 Total long-run supply curve for a commodity produced as a by-product and a
main product: (a) by-product and main production supply curve; (b) total
supply curve.

and so excludes material that ends up as new scrap and is recycled, then the total
supply curve should also exclude new scrap. Conversely, if the total demand
curve includes the demand for all of a commodity, including that which ends up
as new scrap, then the total supply curve should include new scrap as well.

Production Costs and Supply Curves

So far our analysis of supply curves has been totally conceptual or theoretical. We
have described the general shapes of these curves, including situations where the
supply curve is likely to be fairly flat or horizontal (for example, when the

short-run individual product output is well below existing capacity), when it is likely to be vertical or quite steep (for example, when individual product production in the short run approaches existing capacity, when by-product output approaches the by-product constraint, or when secondary production from new scrap approaches the availability-of-new-scrap constraint).

This section explores the efforts of mineral economists and industry analysts to estimate the actual shape of mineral commodity supply curves, particularly those reflecting individual and main product production, on the basis of production costs. It begins by examining the theoretical foundation for these efforts and then turns to comparative cost analysis. The latter entails the collection of mine production costs and the use of this information to estimate both short-run and long-run supply curves.

Microeconomic Theory

The theory of the firm in microeconomics normally differentiates between the short run and the long run. As noted earlier, the short run covers a period of time sufficiently short that some inputs that firms employ, such as capital in the form of plant and equipment, are fixed. In contrast, in the long run all inputs including plant and equipment are variable in the sense that they can be increased or decreased.

In the short run, economic theory defines three cost curves for a firm, or more appropriately for our purposes, for a mine. The first is the *average total cost curve* (ATC), which shows how total costs of production per unit of output vary with output in the short run. Normally, one assumes this curve is U-shaped, as shown in Figure 3.8, initially falling and then rising as output increases as a result of economies and diseconomies of scale. The second is the *average variable cost curve* (AVC), which shows how the variable costs (that is, the costs incurred for inputs that can be changed in the short run) per unit of output vary with output. One normally assumes that it too is U-shaped, as drawn in Figure 3.8. The third is the short-run *marginal cost curve* (MC), which shows how the cost of producing an additional unit of output varies with output. As shown in Figure 3.8, this curve is also usually assumed to be U-shaped. It intersects both the average total cost curve and the average variable cost curve at their minimum points. The reason for this is that when the costs of producing an additional unit are greater than average costs (variable or total), marginal costs are above average costs and so average costs must be rising. And, when marginal costs are below average costs, just the opposite is the case.

Competitive firms as normally defined in microeconomics have no control over the price they receive. Unlike firms with market power, they cannot raise the market price by restricting the amount they offer for sale. This means they face a horizontal demand curve at the market price (P_m), as shown in Figure 3.8.

We also know from microeconomics that competitive firms that maximize their profits have an incentive to remain in production over the short run as long as the market price is above their average variable costs (since they have to pay their fixed costs whether or not they are producing). In addition, they have an incentive

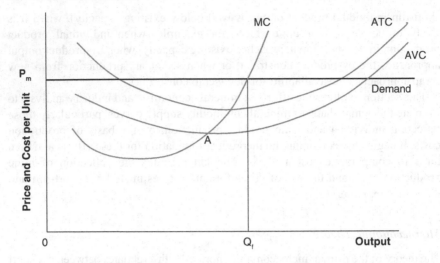

Figure 3.8 Hypothetical average total cost, average variable cost, and marginal cost curves
for a firm.

to increase their output as long as the price they receive for an additional unit of
output is above the costs of producing it. This means that the short-run supply
curve for a profit-maximizing competitive firm is its marginal cost curve above
the price at which it intersects its average variable cost curve.

Thus, in Figure 3.8, the firm's supply curve is its marginal cost curve (MC)
above and to the right of point A (the minimum point on its average variable cost
curve). At the prevailing market price (P_m), this firm will produce and supply to
the market the output $0Q_f$.

Now, assuming the total supply flowing to the market comes from similar
profit-maximizing competitive firms, the market supply curve can be derived by
simply adding (horizontally) the individual firm supply curves for all producers.
So, in theory at least, one could empirically estimate a short-run market supply
curve by measuring the minimum average variable costs and the marginal costs of
all producers.

A long-run market supply curve can be constructed in a similar manner. Of
course, over the long run, all inputs including plant and equipment are variable.
This means that the average variable cost curve and the average total cost curve
are the same. So, normally, economists when discussing the long run talk only
about the latter. The long-run average marginal cost curve again measures the
costs of producing one additional unit, but now it reflects those costs only after the
firm has fully adjusted all its inputs and its production process to the change in
price. A firm's long-run supply curve is its long-run marginal cost curve above the
point at which its average total cost curve reaches its minimum. Again assuming
all suppliers are similar in the sense they are profit-maximizing competitive
producers, we can derive the long-run market supply curve by summing
horizontally the supply curves for all the individual producers.

Comparative Cost Analysis

Reliable data for the marginal production costs do not exist for individual mines or for mining companies as a whole. What is available at the level of individual mines for aluminum, copper, and various other commodities is information on their capacities and their unit costs of production.[2] A few mining companies and consulting firms have used this information to construct comparative cost curves, which under certain assumptions—assumptions we will examine—can approximate market supply curves.

There are different types of comparative cost curves, each based on a different definition of average production costs. However, all of them have costs and price on the vertical axis and capacity on the horizontal axis. They begin at the origin with the lowest cost mine, as shown in Figure 3.9. In this figure, Mine A is the lowest cost mine with unit costs of $0C_a$ and capacity of $0Q_a$. The second lowest cost mine, Mine B, appears next with unit costs of $0C_b$ and capacity of Q_aQ_b. In a similar manner, all the remaining mines in the industry are arranged in order of their costs until finally the highest cost mine, Mine H, is reached. After this mine, there is no more available capacity and the comparative cost curve turns vertical.

The first type of comparative cost curve is based solely on direct costs, or what industry analysts commonly refer to as C1 costs. These are the cash costs (or out-of-pocket costs) incurred for mining, milling and concentrating, smelting and refining, and concentrate transportation and marketing. These costs include expenditures for labor, energy, materials, on-site administration, off-site services essential to the operation of the mine, and property and severance taxes not related to profits. In short, C1 cash costs approximate what economists refer to as average variable costs.

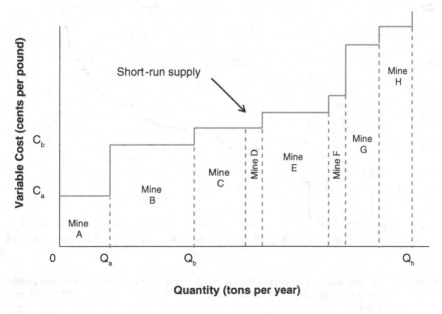

Figure 3.9 Hypothetical comparative cost curve and the short-run supply curve.

Figure 3.10 provides an example of actual comparative cost curves based on data from the consulting firm Wood Mackenzie. It comes from a presentation by a recent CEO of Codelco (Hernández, 2012) and indicates the C1 costs for all the world's significant copper mines for the years 1985, 1990, 1995, 2000, 2005, 2010, and 2011. Any point on these curves shows the millions of tons of mine capacity (measured in terms of copper content) with cash costs at or below the indicated costs in cents per pound.

The figures for C1 costs, it should be noted, are after by-product credits. This explains why a few mines have negative cash costs. In these cases, the value of the gold, silver, molybdenum, and other by-products produced with the copper more than cover the C1 costs of production, including the copper production.

Over time the seven curves shown in Figure 3.10 shift to the right, indicating that copper mine capacity has grown. Perhaps of more interest is the fact that the curves shift up and down over time with copper price movements. During the

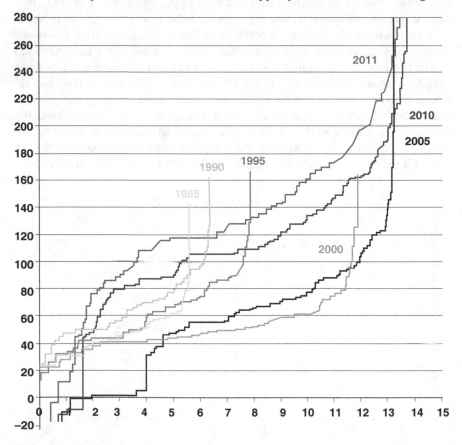

Figure 3.10 Comparative cost curves for the world's copper mines in US cents per pound and millions of tons of capacity for 1985, 1990, 1995, 2000, 2005, 2010, and 2011.

Source: Wood Mackenzie and Codelco as reported in Hernández (2012, p. 15).

latter half of the 1980s, as real copper prices rose, the comparative cost curve shifted upward. During the 1990s as prices declined, the curve moved downward. Then, after 2000, as prices surged, the curve resumed its upward march. All of which suggests that cash or C1 costs tend to rise when prices are increasing and the pressure to expand output is strong. Just the opposite is the case when prices are declining, increasing the pressure to reduce costs.

In addition to C1 costs, industry analysts consider two other measures of average production costs. C2 costs are C1 costs plus depreciation. C3 costs are C2 costs plus interest and indirect costs (such as corporate overhead, exploration costs, research and development costs, and royalties but not income taxes). C3 costs roughly approximate what economists call average total costs since depreciation and interest provide some assessment of the fixed costs associated with plant and equipment.

Mineral economists sometimes assume that a comparative cost curve based on C1 cash costs reflects the short-run market supply curve, but this is the case only when the following, fairly restrictive conditions hold:

- All producers are competitive in the sense that they do not possess the ability and the incentive to influence the market price by limiting their output. When this is not the case and one or a few firms set a producer price, the market supply curve is likely to lie above the supply curve based on C1 cash costs over all or almost all output.
- The short-run marginal costs of mines remain constant as output expands until capacity is reached. As a result, their marginal cost curves are horizontal at the level of their average variable costs (approximated by C1 cash costs) and thus coincide with their average variable cost curves.
- Mining companies close mines when the market price drops below their cash costs (that is, below their marginal costs and average variable costs) and reopen them when the price rises above their cash costs. In practice, we know this is not always the case. Since there are costs associated with closing and reopening mines, companies may not immediately shut down a mine when price drops below its cash costs, particularly if they expect the price to recover in the near future. Moreover, since companies have different expectations regarding future prices, some relatively low cost mines may shut down when the price falls, while higher cost mines remain in operation. Also, mines often are located in remote regions where few alternative employment opportunities exist for workers. As a result, the government may pressure companies, or provide financial incentives, to keep their mines operating even when the market price is below their cash costs.
- The comparative cost curve is stable over time. Again, we know that this condition often is not fulfilled. As Tilton and Landsberg (1999) and other studies show, mines often reduce their costs when the market price falls in order to remain profitable and to avoid closure. Conversely, high prices often make it profitable to exploit lower-grade and more costly ores. They also may reduce the pressure on companies to control costs. As a result, as Figure 3.10

shows, the comparative cost curve may shift up and down as the market price rises and falls. This means that some mines may remain in production due to their cost-cutting efforts even though the market price falls below their erstwhile cash costs.

As a result of these rather restrictive conditions, one should use comparative cost curves to estimate short-run supply curves with some caution. Nevertheless, they can provide useful insights into the nature and shape of short-run supply curves.

The focus so far has been on short-run commodity supply curves. Comparative cost curves can also be of use in assessing the nature of long-run supply curves, though this requires two important changes in their construction. First, the curves should now reflect average total costs (as a proxy for long-run marginal costs). This means that C3 costs rather than C1 costs should be employed in their construction. Second, in addition to all existing mines, the cost curves should include potential mines (since in the long run new mines can be constructed). This, in turn, requires estimates for the average production costs and annual capacity for potential mines, which as they are not yet in existence can at best only be roughly approximated.

In addition, for these curves to portray closely the long-run supply curve, a set of conditions similar to those just discussed for the short run must be met. In particular, all suppliers must be competitive, profit-maximizing producers, and their costs must be stable over the long run.

This latter requirement raises an interesting issue, since we know over the long run the depletion of high-quality, low-cost deposits is likely to push average production costs and hence the comparative cost curve upward, while new cost-reducing technology is likely to have the opposite effect. As we will see in Chapter 9, over the past century for many mineral commodities the cost-reducing effects of new technology have more than offset the cost-increasing effects of depletion, causing real production costs and prices to fall. While this favorable situation may not continue indefinitely in the future, it is clear that comparative cost curves over the long run are much more likely to shift upward or downward than to remain stable.

As a result, comparative cost curves reflect long-run supply curves only at a particular point in time (namely, the period for which they are estimated). They provide an indication of how much suppliers will offer to the market at any particular price after they have had sufficient time to adjust their production process fully to the new price on the assumption that all other current conditions influencing supply remain unchanged. Of course, over the long run, other conditions do change. So the long-run supply curves approximated on the basis of comparative cost curves provide an indication of the long-run supply towards which the market is now moving in response to a change in price. However, long before this targeted supply is reached, changing conditions will cause the market to change course and pursue another long-run supply target.

Despite such caveats, comparative cost analysis can provide useful information about the nature and shape of both short-run and long-run supply curves. Before leaving this topic, it is worth noting a couple of other common uses of comparative cost analysis.

A second use, after the estimation of supply curves, involves the evaluation of investment projects. In fact, this use provided the initial motivation for estimating comparative cost curves back in the 1970s and 1980s. At that time, net present value and internal rate of return techniques were widely employed in assessing the financial attractiveness of developing new mines, expanding existing capacity, acquiring facilities through mergers and acquisitions, and other investment opportunities. These techniques, however, required forecasts of future revenues and expenditures, which due to the volatility of mineral commodity prices and other factors are quite unreliable.

In a fascinating case study of two investment projects—the Toquepala copper mine in Peru and the Bougainville copper mine in Papua New Guinea—Mikesell (1975) carried out a retrospective analysis of their revenues, expenditures, and profits. He then compared the actual results with the projections made in their feasibility studies. The feasibility studies anticipated that the projects would be profitable but only marginally so. In both cases, the feasibility studies significantly underestimated actual costs, but fortunately they underestimated actual revenues by much greater amounts. Both projects as a result turned out to be far more profitable than their feasibility studies anticipated. Of course, other investment projects were not so fortunate and incorrect forecasts, particularly for price, in a number of cases produced disastrous results.

Companies investing hundreds of millions even billions of dollars in mineral projects found the uncertainties surrounding future prices, revenues, and expenditures troubling. Comparative cost analysis was welcomed as an additional tool for providing useful information on the viability of new projects. New mines whose production costs fell within the lowest quartile of existing producers would presumably remain in production even if the market price fell substantially, forcing many other producers to shut down their operations.

A third common use of comparative cost analysis entails scenario comparisons. Of interest here is how comparative cost curves change in response to possible developments, such as the imposition of a mining royalty in Australia, a change in the exchange rate between US and Canadian dollars, a rise in the wage rate in China, the closure of a large platinum mine in South Africa, or the discovery of new, low-cost lithium deposits in Argentina. Comparative cost analysis is used to assess the likely implications of such developments for production costs, supply availability, and the competitiveness of individual producers.

Highlights

The nature of supply varies greatly from one mineral commodity to another. Still, like demand, mineral economists have certain preconceived notions or conventional wisdom with which they start their analyses of supply.

We know, for example, that production entails four steps or stages: exploration and the creation of identified resources; the conversion of resources to reserves;

investment and the development of new mines and processing facilities; and finally the actual mining and processing of ores. Then, once mineral commodities are produced and used, they may be reused thanks to recycling and secondary production.

As a result, commodity supply comes largely from two principal sources—primary production and the mining of mineral deposits, and secondary production and the recycling of scrap. Primary production in turn encompasses both individual products and joint products (main products, co-products, and by-products). Secondary production comes from the processing of new and old scrap. Depending on the purpose of the analysis, however, it may be appropriate to exclude secondary production from new scrap when calculating supply since such scrap arises in the fabrication of final products and so is not an independent and additional source of supply.

We also know that as a result of cost considerations the supply of secondary from old scrap for many commodities comes largely from old scrap flows rather than old scrap stocks. While the estimated quantities of old scrap stocks are for many mineral commodities huge, the costs of recycling these stocks are for the most part simply too high relative to other sources of supply to make them attractive to recycle.

The determinants of supply typically include a commodity's own price and the costs of labor, energy, and other inputs used in its production. In the short run, the constraint imposed by existing mine and processing capacity is important, along with strikes, accidents, and natural disasters. Over the long run, changes in technology, government taxation and regulations, and market structure may have an important impact on supply.

In the short run, mineral economists typically assume that the elasticity of supply with respect to price is inelastic since output is constrained by existing capacity. While this is often the case, short-run supply can be quite elastic when output is low and unused capacity available. Over the long run, since new resources can be found, new reserves created, and new capacity built, the supply of individual and main products tends to be elastic. Indeed, given the likely number of marginal deposits (both known and unknown), the long-run supply curves for many mineral commodities typically rise at first but then level off.

This is not the case, however, for the long-run curves for by-products and secondary supply. Eventually by-product and secondary supply are constrained by the amount of the by-product found in the extracted ore of the associated main product and by the amount of scrap available for recycling. So their supply curves even in the long run turn upward and become vertical. All of which has important implications for the long-run availability of mineral commodities, such as indium, that are currently produced solely as by-products, a concern that Chapter 9 will examine.

Over the past several decades, mining companies and consulting firms have compiled data on the costs and capacities of existing mines in a number of the more important commodity industries. With these data, they have constructed comparative cost curves. These efforts can provide interesting empirical insights into the nature and actual shape of short-run and long-run supply curves. However, comparative cost analysis provides reliable approximations of actual supply curves only when a number of important conditions hold.

Notes

1   Copper ore often contains valuable by-products, such as molybdenum, gold, and silver. The copper equivalent grade of a deposit indicates the percentage of copper that a deposit would have to have if it had no by-products in order to equal the value of its copper and by-products.
2   For many commodities, such as lithium and iodine, this information unfortunately is unreliable or non-existent.

References

Fisher, F.M., Cootner, P.H., and Baily, M.N., 1972. An econometric model of the world copper industry. *Bell Journal of Economics and Management Science*, 3 (2), 568–609.
Gómez, F., Guzmán, J.I., and Tilton, J.E., 2007. Copper recycling and scrap availability. *Resources Policy*, 32, 183–190.
Hernández, D., 2012. Desafíos y oportunidades de la Minería en America Latina, presentation to Expomin 2012, Codelco, Santiago, Chile.
Mikesell, R.F., 1975. *Foreign Investment in Copper Mining: Case Studies of Mines in Peru and Papua New Guinea*, Resources for the Future, Washington, DC.
Tilton, J.E. and Landsberg, H.H., 1999. Innovation, productivity growth, and the survival of the U.S. copper industry, in Simpson, R.D., ed., *Productivity in Natural Resource Industries*, Resources for the Future, Washington, DC, 109–139.
US Geological Survey, annual. *Mineral Commodity Summaries*. Available at: http://minerals.usgs.gov/minerals/pubs/commodity/zinc.

4 Markets and Prices

This chapter explores the behavior of mineral commodity markets and prices. Among the important issues and questions it addresses are: Why are commodity prices so volatile over the short run? Despite the persistent tendency of mineral depletion to push production costs up, why have real prices often declined over long periods? Do investors and speculators accentuate or mitigate price volatility? Do they keep prices higher than they otherwise would be over the longer run? Who are the important players or participants on commodity markets? Are most markets competitive? What exactly is a competitive market? How have mineral commodity markets evolved over time, and are they becoming more or less competitive?

The chapter focuses first on commodity markets, then prices, and finally investor demand and speculation.

Markets

Markets, as Chapter 2 notes, are defined by the nature of the product (so as to include it and all close substitutes) and by the geographic area over which trade can easily occur (so as to encompass all those actors or participants who can buy and sell the product from each other). This section focuses on: (a) the more important types of mineral commodity markets; (b) the variety of market participants; (c) the structure and competitiveness of markets; and (d) the historical evolution of mineral commodity markets over the past half-century.

Market Types

Metals and other mineral commodities are bought and sold in a number of different ways under various institutional arrangements, which in part reflect the size of their markets, the ease with which they can be stored and transported, and the extent to which they are standardized or differentiated. Among the most important types of markets are the following:[1]

1   *Commodity exchanges.* The London Metal Exchange (LME), the New York Mercantile Exchange (Nymex), and the Shanghai Metal Exchange are the three most important and best-known of the commodity exchanges. The LME

trades aluminum, aluminum alloy, copper, tin, nickel, zinc, lead, cobalt, molybdenum, and steel billets. Nymex—including the New York Commodity Exchange (Comex), one of Nymex's two divisions—trades aluminum, copper, gold, silver, platinum, palladium, and uranium. The Shanghai Metal Exchange trades aluminum, copper, lead, nickel, tin, and zinc. Other commodity exchanges based in Chicago, Tokyo, Mumbai, Kuala Lumpur, and elsewhere also trade metals. Prices on this type of market are set on a day-to-day basis (or minute-to-minute basis when the markets are operating) in a manner that balances supply and demand.

Another important characteristic of commodity exchanges is that they offer both spot and futures markets. Participants on spot markets buy and sell physical stocks at a given price for immediate delivery. Participants on futures markets commit to buying or selling a specific amount of a commodity at a given price at some date in the future, such as a month, three months, or a year forward. Futures markets facilitate hedging. A firm that has just purchased copper concentrates for smelting and refining can protect itself from a drop in the copper price over the next three months—the time it may take to process the concentrates and to market the refined copper—by selling short in the three-month futures market. Similarly, a firm already working on a project that will need refined copper in three months can hedge against a price rise by buying three-month copper futures. In addition to hedging, futures markets also facilitate the buying and selling of commodities for investment or speculative purposes. In such instances, market participants are not eliminating the risk of changes in the market price, as with hedging, but instead are assuming this risk in the hope of making a profit. Their efforts, it is worth noting, allows the reallocation of risk to those most willing and able to assume it.

2   *Over the counter (OTC) markets.* In contrast to commodity exchanges, where commodities are traded anonymously through a common mechanism with the exchange itself providing guarantees for participants, OTC markets entail direct trades between participants on a principal-to-principal basis. This means that the risks, including the risk that one of the participants may default, are borne solely by the involved parties. The London Bullion Market Association (LBMA), which trades gold and silver, and the London Platinum and Palladium Market (LPPM) are examples of OTC markets. Like commodity exchanges, prices are set in OTC markets in a manner that ensures supply and demand are in balance.

3   *Producer prices.* In some mineral commodity markets, producers simply announce the price at which they will sell a metal or non-metallic commodity. Normally the number of producers is small and the producer price is the same or nearly the same for all sellers. This type of market tends to reduce, though not eliminate, the volatility of commodity prices over the business cycle. As we will see, however, it can also result in actually physical shortages, requiring producers to ration or allocate their limited supplies, thus forcing consumers at times to simply do without.

Historically, producer pricing has been quite common for both major and minor metals as well as many industrial minerals. For decades, for example, it was the norm for aluminum, nickel, and molybdenum, as well as for copper, lead, and zinc in North America. However, during the 1970s and 1980s, as markets became more global and the number of producers grew, the pricing of these metals shifted to commodity exchanges. It was during this period, for example, that the trading of aluminum and nickel contracts began on the LME. Today some mineral commodities, such as diamonds, platinum group metals, potash, and borates, are still sold on the basis of producer prices.

4   *Negotiated prices*. This type of market often involves bilateral contracts negotiated directly between buyers and sellers without any formal institutional structure. However, where producers are numerous, small, and dispersed, traders may operate as a third party or intermediary, first negotiating purchases from producers and then sales to consumers. Negotiated prices are common with mineral commodities that are differentiated (such as iron ore or coal), are sold in relatively small quantities (such as cadmium or niobium), or are not suitable for trading on commodity exchanges for some other reason. Where prices are negotiated, trade journals such as *Metals Bulletin*, *Platts Metal Week*, and *Industrial Minerals* play a useful role in publishing reference prices based on confidential telephone surveys.

For many years the price of iron ore in the European market was set for the coming year in annual negotiations between the Brazilian producers and the major steel producers in Germany and other European states. Similarly, the price of iron ore in the Asian market was the result of negotiations between the big Australian iron ore producers and the major Japanese steel companies. Since iron ore can be shipped from Australia to Europe or from Brazil to Japan, prices in these two geographic areas tended to move more or less together.

Since the turn of the century, however, this system of pricing iron ore has come under increasing stress, in part the result of China's rise as the world's largest consumer. Even more important is the growing disenchantment of the major producers. With prices fixed for a year and with frequent severe shortages, they have often received prices substantially below the spot prices at which many small Indian and Chinese mines were selling their iron ore. For this reason, BHP Billiton, Rio Tinto, and other major iron ore producers are now selling iron ore under long-term arrangements where the price varies quarterly or monthly over the contract period based on the spot price or some other price index reflecting changes in market conditions.

Similar arrangements allowing price to fluctuate under long-term contracts are common for copper concentrates and other unrefined nonferrous metals. In such cases, after negotiating the processing charges, an independent smelting and refining company may purchase concentrates from a mining company at a price that varies with the LME copper price. Alternatively, the smelting and refining company may agree to process the concentrates for the mining company, with the latter retaining ownership, for a negotiated fee.

Negotiated prices are also found for products produced and consumed by only a few firms, such as the cold-rolled steel sheet used for automobile bodies, the aluminum sheet used for beverage cans, and the tinplate used for food containers. They are also the norm for the bulk ferroalloys such as ferromanganese, for various ferroalloys used to produce stainless steels and specialty steels such as tungsten and cobalt, for many minor metals such as cadmium and bismuth, and for various mineral-based products such as lithium carbonate.

Market Participants

Participants in any particular market include all the individuals and institutions that buy and sell the commodity, regardless of where they actually are within the relevant geographic market. They can be separated into two different groups—those primarily interested in buying, selling, or trading physical material; and those interested in commodities as paper assets or investments.

The first group includes producers, consumers, traders, and governments. Producers participate because they want to sell the metals and other mineral commodities that they mine and process. Consumers participate to obtain the inputs they need to construct automobiles, bridges, cell phones, and other products. Traders buy mineral commodities from various producers and then resell their stocks to consumers elsewhere. They also use markets to hedge price risks and to take advantage of arbitrage opportunities arising from price disparities in different locations or among the spot and futures markets on commodity exchanges. Governments participate when their state-owned enterprises are producers or consumers of mineral commodities. In addition, some governments maintain stockpiles of gold, copper, rare earth minerals, and other mineral commodities for monetary, strategic, and other purposes. They are, of course, buyers and sellers of mineral commodities only when they are building up or selling off their stockpiles.

Investors constitute the second group of market participants. They are interested in buying or selling metals and other commodities because they hope to make a profit in the process. In addition, many investors believe that the returns from commodities are inversely correlated with bonds and equities, and so purchase them to diversify their investment portfolios and increase the stability of their expected returns.

Though there is a long history of investors buying and holding physical material, they have largely operated on futures markets, committing to buying or selling a specific amount of a commodity at a given price at some date in the future, such as a month, three months, or a year forward. Then at or before the due date, they close out their long or short positions, and so never have to actually deliver or take possession of the product. In recent years, however, the amount of physical material held by investors has increased, in large part the result of the growth in exchange trade funds (ETFs) that buy and hold various commodities including gold, copper, and, since 2010, lithium. As a result, investors are increasingly active on spot markets, buying and selling commodities for immediate delivery.

Some mineral economists and market analysts make a distinction between speculators and investors. They define speculators as those who buy commodities on the margin in the hope of making a quick profit. Their demand is quite sensitive or elastic with respect to price. In contrast, they define investors as those who buy commodities to diversify and balance their investment portfolios. As they desire to hold a certain portion of their portfolios in commodities, their demand is not particularly sensitive to changes in prices. Here, however, we will use the two terms—investors and speculators—as synonyms to describe those who are solely interested in commodities as investment opportunities.

Market Structure and Competitiveness

Mineral economists, like economists more generally, differentiate between competitive and non-competitive markets. A competitive market by definition is a market composed of competitive and only competitive actors. A competitive actor is a market participant with no market power. This means that it lacks the ability or the incentive (or both) to control the market price.

Economists have traditionally defined market power as simply the ability to influence the market price with no consideration as to whether or not the firm has an incentive to do so. So a competitive firm that is a seller cannot raise the market price by restricting its sales. Similarly, a competitive firm that is a buyer cannot reduce the market price by limiting its purchases. In the case of mineral commodities, however, this definition can easily lead to the conclusion that the copper, nickel, aluminum, iron ore, and other commodity markets are non-competitive when this may in fact not be so.

To illustrate why, we can consider the global market for refined copper. Codelco is the largest producing firm. Depending on the year, it accounts more-or-less for some 10 percent of world output. Chile is the largest copper-producing country, with roughly one-third of world output. Clearly, if Codelco cut its output or if Chile required its copper companies to limit their sales, this would have an immediate impact on the market price of copper. Over the long run, the impact would be much more modest and could be negligible, as the higher price in the short run would encourage other producing firms and countries to expand their output and consumers to search for ways to reduce their copper needs. As a result, if Codelco cut its production, it might earn higher profits than otherwise for a while. However, over time any extra profits would decline and eventually turn negative as the market price fell back toward its pre-restriction level and the company lost market share to other producers.

Whether restricting output is in the interest of Codelco and its shareholders (the people of Chile) depends on the company's goals and how it discounts future profits. Normally, economists assume that companies are striving to maximize their profits—or more precisely the net present value (NPV) of the stream of their present and future profits (see Box 4.1). In this case, if restricting output increases the NPV of present and future profits, then the company has an incentive to do so. However, restricting output may reduce the NPV of its profits. In this case, although the firm can restrict output and influence the market price for a time, it has no incentive to do so. Instead, it behaves as a competitive firm. If the same is true for BHP Billiton, Anglo American, and other large copper producers, the global market for refined copper is competitive with no firm exercising market power and influencing the market price.

Box 4.1 Present Value and Net Present Value

A dollar received today is worth more than a dollar received a year or ten years in the future. This is in part because the dollar received today can be invested and so can earn a return during the coming year or decade. As a result, it will be worth more than a dollar a year or ten years from now. The concept of present value allows us to determine the worth today of revenues that will be received or expenses that will be incurred in the future. It discounts the future revenue or expense by the time value of money, which we can approximate by the interest on a government bond or some other riskless investment. For example, if the time value of money is 5 percent per year, then the present value of a dollar received one year from now is 95.2 cents (or one dollar divided by 1.05). The present value of a dollar received ten years from now is 61 cents (or a dollar divided by 1.05^{10}). When an investment project generates a stream of net revenues (revenues minus expenses) over a number of years in the future, its net present value is simply the sum of the present values for all its future revenues and expenses.

In practice, showing that firms have the ability to influence the market price is much easier than that they have both the ability and the incentive. For many mineral commodity markets this makes it difficult to determine whether or not they are competitive. For the major metals, despite frequent claims that the largest producing firms (and producing countries) possess market power, the available evidence is far from convincing. When producers restrict their output, they tend to shut down high-cost capacity during periods of low market prices, which is consistent with the behavior we expect of competitive firms when price falls below operating or variable costs.

During 2002 and 2003 BHP Billiton, Codelco, and Phelps Dodge announced a reduction in sales that some maintain was a tacit cooperative effort to raise the market price of copper or alternatively to keep it from falling further. At the time the price was quite depressed. BHP Billiton reduced sales by cutting output, which as just noted, could simply reflect the fact that the market price was below the production costs of the facilities it closed. Codelco, on the other hand, rather than reduce production, built up its inventories. These inventories were then sold a year or so later, after the market price had risen appreciably. So Codelco's behavior is also consistent with an astute competitive producer with sufficient knowledge of the copper market to anticipate a significant rise in price.

At times, the mere existence of producer prices and negotiated prices is thought to reflect the presence of market power. The idea here is that when mineral commodity markets are competitive market participants rely on commodity exchanges or OTC markets to buy and sell. When producers are sufficiently few in number to permit producer pricing or negotiated pricing, then they (and perhaps buyers in the case of negotiated prices) presumably enjoy some influence over the market price.

This argument might be convincing if market power depended solely on the ability to influence price. However, as we have seen, market power depends on both the ability and the incentive, and consequently firms with a large market share may be competitive. Moreover, as Scherer (1980) shows after reviewing the available literature on price leadership (see Box 4.2), producer prices need not reflect market power. Particularly when they change frequently, they may simply move up and down with market conditions, following the competitive price.

On the other hand, it is worth noting that prices set on commodity exchanges are not necessarily competitive. During the 1960s, for example, European zinc-producing firms stabilized the LME zinc price by buying or selling the quantities necessary to keep the price at the level they wanted. The LME tin price was similarly stabilized and at times maintained above its competitive level by the major producing and consuming countries through a series of International Tin Agreements spanning the years 1956 to 1985. These agreements relied upon a buffer stock stabilization fund and export quotas on producers to control the market price.

Box 4.2 Producer Prices and Price Leadership

Economists who have studied the nature of producer prices distinguish between three types of price leadership—dominant firm, collusive, and barometric. *Dominant firm price leadership* occurs in markets where one firm accounts for a large share of total sales and the rest of the market is divided among a large number of smaller fringe firms that sell all their output at the price set by the dominant firm. *Collusive price leadership* occurs in oligopolistic markets with a few large firms, all of which may enjoy some degree of market power. Here, the producer price is likely to be above the price that would prevail in a competitive market. *Barometric price leadership* occurs when the price leader sets prices that closely reflect market conditions and the competitive price, just as a barometer reflects changes in atmospheric pressure. Here, the number of firms is likely to be larger and their market share smaller than in the case of collusive price leadership. In addition, both the price leader and the producer price tend to change with some frequency. For more on the nature of price leadership, see Scherer (1980, pp. 176–184).

Turning from producing companies to producing countries, we again find little evidence to suggest the presence of market power. Countries producing bauxite, copper, iron ore, and other mineral commodities in the late 1970s and early 1980s tried to follow the example of OPEC and the oil-producing countries in stabilizing and raising prices, but were completely unsuccessful. Even today from time to time some suggest that Chile should be the Saudi Arabia of copper, since that country alone accounts for one-third of the world's total output of primary copper.

In response, however, saner voices more familiar with the global copper industry point out that any attempt by Chile to restrict its output and artificially raise the market price of copper would simply encourage an expansion of production in Peru, Indonesia, Canada, Mongolia, Zambia, and other countries. Unlike Saudi Arabia and oil, the cost of producing copper is not all that different in Chile than in the rest of the world.

For these reasons, when confronted with claims that this or that mineral commodity market is non-competitive, we should be cautious. This does not mean that all metal and non-metallic markets are competitive. A cursory review suggests that some are and others are not. The markets for ferrous scrap, for example, typically encompass many buyers and sellers with none having any control over the market price. At the other extreme there are only two producers of niobium. They both may have the ability and incentive to influence the price they receive. The conditions required for the exercise of market power may also exist for other minor metals and non-metallics where production is concentrated in the hands of a few producers.

Even among the major metals—steel, copper, aluminum, nickel, for example— producers appear to have possessed some market power in the past, when their markets were much smaller and more regional. Over the past 50 years, however, a number of historical developments have substantially enhanced the competitiveness of commodity markets. At the end of this section we will examine these developments and consider whether this tendency toward more competitive markets is likely to continue in the future. Before doing so, however, we need to explore the nature of non-competitive markets, and the reasons why we care whether markets are competitive or non-competitive.

Nature of Non-competitive Markets

While competitive markets are homogeneous in the sense that all market participants are competitive, there are many different kinds of non-competitive markets. We can categorize non-competitive markets according to the number of their buyers and sellers, as shown in Table 4.1.

Examples of commodity markets that have just one seller (monopoly, monopoly with oligopsony, and bilateral monopoly) or just one buyer (monopsony, monopsony with oligopoly, and bilateral monopoly) are hard to find today. Some would argue that in the case of the rare earth minerals China was until recently a monopolist or nearly a monopolist with over 95 percent of world production. There are also some examples from the more distant past. The aluminum market

Table 4.1 Market types according to the number of sellers and buyers

	One seller	Few sellers	Many sellers
One buyer	Bilateral monopoly	Monopsony with oligopoly	Monopsony
Few buyers	Monopoly with oligopsony	Bilateral oligopoly	Oligopsony
Many buyers	Monopoly	Oligopoly	Competitive

88   *Mineral Economics*

before World War II was divided into the North American market, where Alcoa was for many years the sole producer, and the European market, where Pechiney and Alusuisse were the only two suppliers. Today, of course, the two markets have long since merged into a global market and the number of producers has multiplied many fold. Similarly, for a time after World War II the uranium industry was a monopsony, with the US government the sole buyer.

Examples of mineral commodity industries with just a few sellers or buyers are easier to identify. As was noted earlier, there are only two significant producers of niobium, providing an example of oligopoly. The platinum group metals, where the number of producers is also quite limited, is another example. Bilateral oligopoly also exists. Cold-rolled steel sheet used to make automobile bodies, aluminum sheet for beverage cans, and tinplate for food containers are all sold by a few producers to a few consumers. Ferrous and nonferrous scrap markets, when regional rather than global in scope, may have many sellers but only a few buyers, and so provide examples of oligopsony.

Another non-competitive market, not shown in Table 4.1, but at times of interest, is the dominant firm type of market. Here, one finds one dominant seller, which sets the market price, and a fringe of small competitive sellers in a market where buyers are numerous. So although there are many sellers, the dominant firm is sufficiently large to possess market power. In modeling the behavior of such markets, economists generally assume that the dominant firm sets the market price by first determining its own demand curve (by subtracting the output that the fringe producers will supply at various prices from the market demand curve) and then setting the price that maximizes its profits, just like a monopolist facing the same demand curve as the dominant firm.

Often the world oil industry is depicted as a dominant firm market with OPEC as a whole or Saudi Arabia (with or without other low-cost Middle East producers) assuming the role of the dominant firm. This model may also apply to metal and non-metallic markets where a few main-product producers face a large number of by-product, co-product, and secondary producers.

Again, however, it is critical to remember that where one or several producers possess a large share of the market this does not necessarily mean they have both the ability and the incentive to influence the market price. If an attempt to raise price artificially stimulates the entry of new firms, the expansion of existing producers, and the efforts of consumers to substitute alternative materials, any increase in profits in the short run may quickly be overwhelmed by a loss of market share and lower profits in the future. So what may at first blush look like an oligopolistic or dominant firm market may in fact be a competitive one.

Equity and Efficiency of Competitive Markets

Competitive markets have a number of socially desirable characteristics. Since market participants have no market power, they cannot earn excess profits by manipulating the market price. For most people, this seems equitable.

In addition, competitive markets promote both production and allocative efficiency. *Production efficiency* reflects *how* goods are produced and ensures that production costs are minimized. In competitive markets price is determined by the costs of the marginal producers. Firms that fail to minimize their costs are likely to find that their costs are higher than the market price. This causes such firms to lose money and eventually to exit the industry.

Allocative efficiency reflects *how much* of various goods and services the economy produces and hence the distribution or allocation among possible uses of society's labor, capital, and other factors of production. In competitive markets firms have an incentive to expand their output up to the point where the market price just equals the costs of producing another unit of output. Since consumers will continue to purchase a good until the benefits to the marginal customer just equal the market price, the benefits the marginal consumer receives from the last unit produced just equal the costs of producing that unit. Unless there are externalities (that is, benefits or costs of producing and consuming a good that third parties incur rather than the producer and consumer, about which much more is said in Chapter 10), competitive markets as a result promote allocative efficiency. In contrast, in markets where producers have market power, they have an incentive to stop production before the costs of producing the next unit reach the market price in order to keep the market price above its competitive level. Where sellers have market power and keep the price below marginal costs, just the opposite is the case.

Production and allocative efficiency are static attributes. *Dynamic efficiency*, in contrast, encompasses benefits arising from the development and diffusion of new technology. Over the longer run, dynamic efficiency is of much greater importance than static efficiency. Whether competitive markets are superior to non-competitive markets in promoting dynamic efficiency is somewhat less clear. A number of the available studies on this topic—including one of our first books (Tilton, 1971), a case study of the creation and then diffusion internationally of the transistor and other early semiconductor technology—concludes that a market with both large and small firms, even though the large firms may possess some market power, may be desirable. Large firms with greater access to capital seem to have an advantage in pioneering new inventions and innovations, while small and especially new firms are particularly effective in promoting their rapid and widespread diffusion.

Despite this possible reservation, public perception and public policy typically prefer competitive to non-competitive markets. This is as true for metals and non-metallic commodities as for other products. As a result, many countries and the European Union have laws fostering competitiveness and government agencies to enforce them.

Historical Evolution

The use of metals and non-metallics in the 1950s was quite modest compared to today. For example, the global consumption of nickel and copper in 1950 was only around 10 and 15 percent, respectively, of recent levels. Moreover, it was largely concentrated in the United States, the one large economy that remained more or less intact after World War II. As the economies of Britain, France, Germany, and other European states recovered from the ravages of the war, consumption spread there as well, and then somewhat later to Japan.

During this period, production was also much more concentrated in a small number of firms, many located in the United States and Canada, with others in Mexico, Chile, South Africa, and other developing countries. International trade, particularly in low-cost bulk commodities such as bauxite, manganese, and iron ore, was constrained by the high cost of shipping. The world was also divided, with the United States, Western Europe, and other countries with market economies in one camp (often referred to as the West), and the Soviet Union, Eastern Europe, China, and other countries with communist governments and central planning in the other (often referred to as the East). This second group strove for self-sufficiency in mineral commodities and other goods, and only resorted to imports, particularly imports from the West, when their own production could not satisfy domestic demand. In the centrally planned countries, state enterprises did the mining and processing of mineral commodities, while the market-economy countries mostly relied upon private companies.

Against this backdrop several major developments took place over the next half-century that greatly reshaped the global mineral sector:

1   The first of these, confined to the West during the 1960s and 1970s, was the nationalization of many of the world's metal mines and processing facilities. Most, though not all, of these conversions from private to state ownership took place in developing countries, such as Jamaica, Zambia, Bolivia, and Chile. The world oil industry witnessed a similar though more complete and somewhat earlier trend toward state ownership.[2]

    Driving nationalization was the belief that international mining companies were capturing the lion's share of the profits or rents from inherently valuable mineral resources, leaving the host countries that owned these resources with just a small share of the benefits. Of even greater importance was the desire in many developing countries to ensure that foreign companies did not control major sectors of their economy (sectors that provided much of their foreign exchange and government revenues), thereby compromising their political independence. For many of these countries, independence had only recently been won after a long and difficult struggle.

    The tide toward state ownership crested in the 1980s, as the inefficiency and other problems associated with state enterprises became painfully clear in Latin America, Africa, and elsewhere. Since then, many countries have privatized erstwhile state mining companies, and opened up their mineral sectors to international mining companies and other private interests for exploration and development.

    The wave of privatization—some are now suggesting—may be over, as Bolivia, Ecuador, Venezuela, and a few other countries take steps to control private mining companies within their territories. At this point, however, it is too early to know for certain whether these actions are permanent, let alone whether they reflect a new global trend.

2　A second major development, again confined to the West, occurred during the 1970s and 1980s. Many international oil companies, flush with cash from the high price of oil, entered the mining industry, often by purchasing large mining companies, such as Anaconda and Kennecott. These companies invested heavily in their new acquisitions, helping the declining US mining industry recover during the 1980s. The marriage between energy and mineral companies, however, was not a happy one, and by the 1990s had largely ended in divorce. The cultures and expectations of the two groups of firms were quite different; in addition, the investments made by oil companies in mining were slow to turn a profit. The entry and exit of the oil companies, however, left their mark on the mineral sector, particularly in terms of the injection of substantial new capital that both raised productivity and expanded capacity.

3　The collapse of the Soviet Union in 1989 and the shift in China from a centrally planned to a market-based economy during the 1980s and early 1990s led to global markets for most mineral commodities that no longer were partitioned into two major geographic blocs. These developments also reinforced the trend toward the privatization of the mineral-producing operations that occurred in the West during the 1990s. This was particularly so in Eastern Europe, where Poland, the Czech Republic, Hungary, and other countries opened their economics, including the mineral sector, to private enterprise.

4　The revolution in ocean shipping over the past 60 years with respect to low-value bulk commodities is yet another important historical development. As Lundgren (1996) points out (see Box 4.3), the costs of ocean transportation over the past several centuries has declined dramatically for different types of products at different times.

　The most recent burst of technological change has transformed the shipping of petroleum, coal, iron ore, magnesium, bauxite, and other bulk mineral commodities. Thanks largely to this development, the iron ore market is no longer divided into two geographic areas. While Brazil mostly serves the major European steel producers, and Australia the steel producers in China and Japan (with South Africa well positioned to serve either or both areas), prices in the two markets move more or less together. Brazil can now ship to the Far East if prices there are higher than in Europe, and similarly Australian iron ore can be sold in Europe when the opposite is the case.

Box 4.3 Three Revolutions in Ocean Transportation

Three distinct technological revolutions have dramatically cut the costs of ocean transportation over the past 600 years.

The first revolution started with the wooden, three-mast sailing ships of the fifteenth century and evolved over several centuries, ultimately producing the efficient, wooden, square-rigged sailing vessels of the early 1800s. It fostered an explosion in international trade in commodities with high values relative to their weight—spices, gold and silver, fine china and other ceramics, and woolens. It led to ever-greater global economic integration, particularly between Europe on one hand and Asia, Africa, and the Americas on the other.

The second revolution began in the second half of the nineteenth century and continued until about 1910. Among other developments, it witnessed the shift from wooden sailing ships to metal vessels powered by steam engines and coal. This second revolution greatly reduced the costs of transporting commodities of moderate value. It made wheat and other grains from North America and the Ukraine competitive in European markets, causing France and other countries even to this day to subsidize and protect their domestic farmers.

The third revolution took place after World War II. It produced various advances, including the shift from the steam engine to the internal combustion engine and in particular to the diesel engine, whose shaft was coupled directly to the screw propeller. This era also saw the development of specialized bulk carriers of greater and greater size. In the late 1950s, for example, ships of 40,000 dwt (deadweight tons) or less accounted for almost 100 percent of the seaborne trade in iron ore. By the end of the century this figure had dropped to about 5 percent—by then ships of 100,000 dwt or larger were the norm.

The third revolution had a major impact on mineral commodity markets. It dramatically reduced the costs of shipping iron ore, coal, oil, bauxite, manganese, phosphate, copper ore and concentrate, and other low-value bulk commodities. The costs of shipping a ton of iron ore from Brazil to Europe, for example, dropped from about 30 to 5 dollars between 1960 and the end of the century. As a result, iron ore producers in Sweden and elsewhere in Europe found it increasingly difficult to compete even in their own backyard. More and more markets for such low-value bulk commodities became global, with producers in Brazil, Australia, and other remote areas shipping their output thousands of miles to distant markets in Europe, Japan, China, and North America.

Source: Lundgren, 1996.

---

5   The explosion in commodity consumption over the past 50 years has and still is reshaping the mineral sector. Global consumption of all the major metals—aluminum, copper, iron ore, nickel—has more than tripled over this period. The use of potash, lithium, and other non-metallics has also grown dramatically. This growth has facilitated an increase in the number of producing firms and countries, and diversified the mineral industries geographically. In addition, as Chapter 2 points out, for decades mineral commodities were largely consumed in the United States, Western Europe,

Japan, and other developed countries. Over the past decade, however, in an important and dramatic shift, China and other developing countries have become important consumers. Indeed, China is today the world's largest consumer of iron ore, aluminum, copper, nickel, potash, asbestos, salt, and many other mineral commodities.

6   Over the past 50 years the international mining community has witnessed numerous waves of mergers and acquisitions. Today the largest international mining companies—BHP Billiton, Rio Tinto, Anglo American, Vale, Glencore, Freeport McMoRan—are all the products of earlier combinations.

Mergers and acquisitions create both benefits and concerns. They reduce costs, not so much by realizing economies of scale as these are mostly a function of the size of existing mines, which mergers and acquisitions do not change, but rather by reducing overhead and the cost of acquiring materials and other supplies. Such savings economists refer to as economies of scope. The concerns arise, as Chapter 7 assesses in greater detail, because mergers and acquisitions, particularly those over the past decade, increase market concentration and fears of market power.

BHP Billiton's attempt to acquire Rio Tinto in 2007 and 2008, for example, raised such fears in the Far East, as together these two companies supply the lion's share of the iron ore going to Chinese and Japanese steel mills. Even after the proposed merger fell through, the plans of the two companies to integrate their iron ore operations in Australia caused concern in China and other countries.

In looking at the evolution of mineral commodity markets since 1950, it is worth noting that most of the important developments just described have increased competitiveness. These include the dramatic fall in the costs of transporting iron ore and other mineral commodities, the emergence of new producers particularly in developing countries, the dramatic growth in consumption and the size of mineral commodity markets, the development of truly global markets due to the disintegration of the Soviet Union, and the trend in former Soviet states as well as China to turn from central planning to market economies (and from autarky to open economies encouraging imports and exports). These developments have more than overwhelmed any negative effects on competition flowing from the various waves of mergers and acquisitions in the past. Many mineral commodity markets as a result have become more competitive, and some—such as nickel, aluminum, and in North America copper—have moved from producer pricing to the use of commodity exchanges.

Prices

As Figure 4.1 illustrates for aluminum, copper, iron ore, lead, nickel, phosphate rock, potassium chloride, and zinc, the prices for many mineral commodities display two interesting characteristics. First, they fluctuate greatly over the short run. It is not unusual for prices to double or fall by half within a year or two. This volatility creates problems not just for producing firms and countries, but for consumers as well. It has led to numerous efforts to stabilize commodity markets, which for the most part have had limited success.

Second, over the long run real prices follow secular trends. For many mineral commodities, the long-run trends over the past four decades have been downward until at least 2003 or so. Figure 4.1 highlights this secular decline by showing the linear trend in real prices over the 1970–2003 period for each of the commodities it considers.[3]

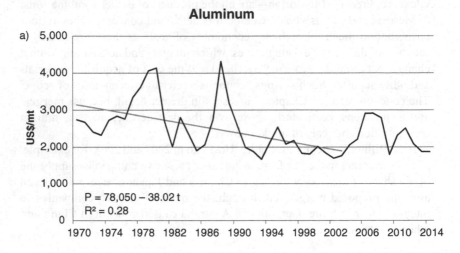

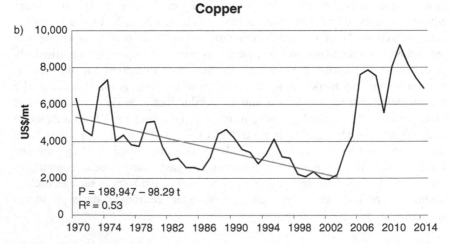

Figure 4.1 Average annual real prices for aluminum, copper, iron ore, nickel, lead, phosphate rock, potassium chloride, and zinc, 1970 to 2014 in 2014 dollars per metric ton (sources: UNCTAD (annual); World Bank (annual)).

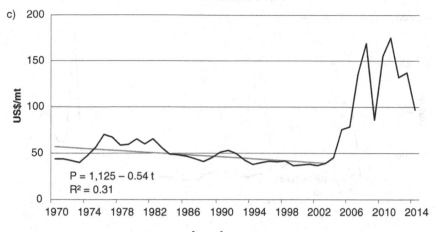

Iron ore

c)

$P = 1{,}125 - 0.54\,t$
$R^2 = 0.31$

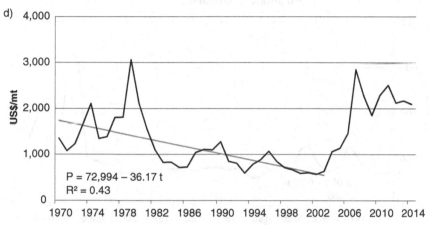

Lead

d)

$P = 72{,}994 - 36.17\,t$
$R^2 = 0.43$

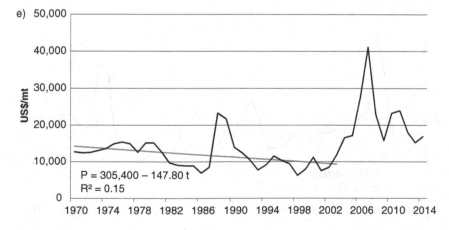

Nickel

e)

$P = 305{,}400 - 147.80\,t$
$R^2 = 0.15$

Figure 4.1 continued

Phosphate Rock

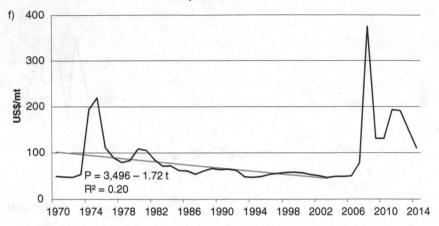

f)

$P = 3{,}496 - 1.72\,t$
$R^2 = 0.20$

Potassium Chloride

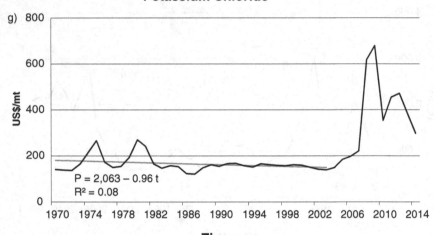

g)

$P = 2{,}063 - 0.96\,t$
$R^2 = 0.08$

Zinc

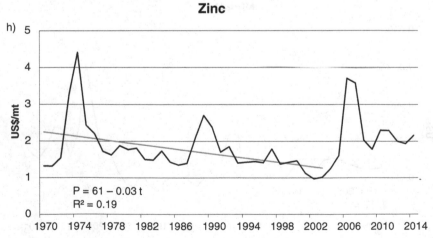

h)

$P = 61 - 0.03\,t$
$R^2 = 0.19$

Figure 4.1 continued

Since 2003, despite sharp drops in 2008 and early 2009 associated with the global recession at that time, real prices for many mineral materials have risen, sharply in some cases such as copper. The strength and persistence of the recent rise in prices have led many to conclude that the declining long-run trends for many mineral commodities bottomed out in the early 2000s and are now rising.

This section focuses on the causes of price changes over both the short and long runs. Simple economics suggests that, in the case of competitive markets at least, it is all a matter of supply and demand. When demand is rising faster than supply, prices will rise and vice versa. This does not mean, however, that supply and demand necessarily play equally important roles. Indeed, as we will see, shifts in demand are largely behind changes in mineral commodity prices in the short run, at least when production is at or near capacity, while changes in supply dominate the long-run trends.

Short-Run Volatility

Mineral commodity markets have long been known for their instability, for their feast-or-famine nature. One of the driving forces behind the multiple mergers in the American steel industry at the dawn of the twentieth century that created the US Steel Corporation in 1901 (which at that time possessed some two-thirds of the country's steel-making capacity) was the desire to control the volatile steel market. Fluctuations in steel prices during the 1880s and 1890s inflicted great pain on both producers and consumers.

A highly concentrated market structure where one or a few companies dominate the market and set a producer price may reduce price fluctuations but it does not eliminate the problems associated with market volatility or their underlying causes. This is because the three characteristics of short-run mineral commodity supply and demand responsible for market instability persist no matter how concentrated the market may be.

First, as output approaches the capacity constraint, market supply becomes increasingly price inelastic. In a competitive market, as shown in Chapter 3, the short-run supply curve turns upward and above some price becomes vertical. In a producer market, the curve simply ends (and implicitly becomes vertical) when producing firms no longer have sufficient supply to satisfy additional demand at the producer price.

Second, for the reasons discussed in Chapter 2, mineral commodity demand also tends to be inelastic with respect to price in the short run. So the slope of the demand curve is quite steep.

Third, demand tends to be highly elastic to changes in national income over the business cycle. As Chapter 2 pointed out, this is because most mineral commodities are largely used in four end-use sectors—construction, capital equipment, transportation, and consumer durables—whose outputs are particularly sensitive to the business cycle. During a recession, these sectors contract much more than the economy as a whole. During a boom, their sales soar. As a result, the demand curves for most mineral commodities experience sizable shifts over the business cycle.

These attributes of short-run supply and demand are shown in Figure 4.2a for mineral commodities sold in a competitive market and in Figure 4.2b for those sold in a producer market. In both instances, we have assumed that supply comes from individual or main product production (see Chapter 3). This simplifies the analysis, but changing this assumption would not alter the conclusions, since total supply, regardless of the combination of sources from which it comes, is in the short run at some point constrained by the available production capacity. As supply approaches this constraint, it becomes inelastic with respect to price.

The two characteristics of mineral commodity demand over the short run—its low elasticity with respect to price and its high elasticity with respect to income—are reflected in Figures 4.2a and 4.2b by the steep slopes of the demand curves and by the shifts in the demand curve over the business cycle. The curve D_r reflects

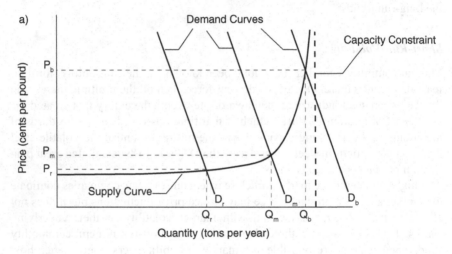

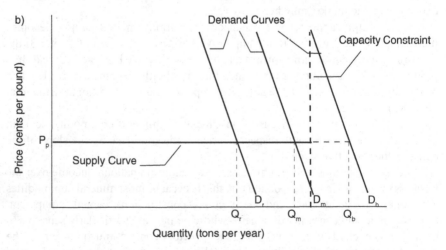

Figure 4.2 Short-run price fluctuations caused by shifts in the demand curve over the business cycle: (a) competitive market; (b) producer market.

demand during a recession, the curve D_m demand at the midpoint of the business cycle, and the curve D_b demand when the economy is booming.

Prices for mineral commodities sold on competitive markets vary greatly over the business cycle. In Figure 4.2a they range from a low of P_r during recessions to a high of P_b during economic booms. The output that producers supply to the market also changes markedly, from a low of Q_r to a high of Q_b. At the low output, producers are forced to either idle production capacity or to add to their inventories. At the high output, they are pushing the limits of their available capacity.

This explanation for the volatility of mineral commodity prices assumes that over the business cycle production is constrained by existing capacity. If this is not the case, if sufficient excess capacity exists so that even during economic booms production is not constrained by capacity, then the demand curve will always intersect the supply on its relatively flat or elastic segment. In this situation, short-run price volatility is likely to be less and to arise largely because of shifts in the supply curve rather than the demand curve. The supply curve, for example, may shift down during a recession—causing price to fall even further—as producing firms struggle to find ways to reduce their costs.

While both mineral and agricultural commodities are known for their price volatility over the short run, there is an important difference between the two. Fluctuations in metal and other mineral commodity prices, as we have seen, arise largely from shifts in the demand curve when production is constrained by existing capacity during economic booms. Of course, the supply curve can shift as well over the short run. Mine accidents, equipment failures, strikes, and other production problems can cause the supply curve to move up and to the left, while the inauguration of new capacity already under construction can cause the curve to move down and to the right. And, as just noted, the supply curve may shift down and to the right during recessions, as a result of producers' efforts to cut costs. However, such shifts are normally less important than shifts in the demand curve.

In the case of agricultural products, demand varies much less over the business cycle as people need wool and cotton for clothes and wheat and bananas for food, regardless of whether the economy is stagnant or booming. Instead, it is shifts in the supply curve—the result of crop failures due to droughts, freezing weather, pests, disease, floods, and other disasters—that cause price volatility. As a result, when output is down, prices are likely to be up, which dampens the fluctuations in total revenues and profits for the industry as a whole (though obviously not for those particular producers directly impacted).

Unfortunately for mineral producers, output and price move together. When prices are down, so are output and sales. This means that total revenue and particularly profits are extremely volatile. When times are good for mineral producers they are very good, and when they are bad they are very bad.

So far we have considered only commodities sold on competitive markets. The situation is somewhat different in the case of producer markets. Obviously, if firms faithfully adhere to one producer price over the business cycle, there is no price volatility. Even where some open or secret discounting occurs, where the producer price is adjusted upward during booms and downward during recessions, price instability will generally be less than in competitive markets.

However, the three characteristics of short-run supply and demand responsible for the volatility of competitive mineral commodity markets are just as prevalent for mineral commodities sold on producer markets. The market instability they produce simply manifests itself in other ways. With producer markets, for example, the burden of keeping supply and demand in balance over the business cycle falls much more on supply, forcing firms to vary their output over the short run far more than is the case with competitive markets. In addition, producer markets can suffer from actual physical shortages, where the available supply is unable to satisfy the demand at the prevailing producer price. This situation is illustrated in Figure 4.2b. When the economy is booming, the quantity demanded is Q_b, while the maximum amount the industry can supply is limited to its capacity. This physical shortage—the difference between Q_b and capacity—forces producers to allocate or ration their insufficient supplies to customers.

Moreover, firms in producer markets still suffer from sizable fluctuations in total revenue and profits over the business cycle as a result of the instability in demand and the impact that this has on sales. In this respect, the adverse effects of market instability are similar for firms in both producer and competitive markets.

Long-Run Trends

For the reasons Chapter 3 examines, the long-run supply curve for most mineral commodities rises at first and then levels off. Whether the slope eventually becomes horizontal or, as Figure 4.3a suggests, rises modestly is uncertain. However, if the relatively flat portion of the supply curve covers the relevant range of future global demand, whether it is nearly or completely horizontal matters little. In either case, demand has little influence on long-run prices. As Figure 4.3a illustrates, if demand grows slowly over the next 20 years, causing the demand curve to shift only from D_0 to D_1, the price at which the long-run supply curve (S_0) intersects the demand curve (D_1) is P_1. On the other hand, if demand grows rapidly over the next 20 years, causing the demand curve to shift to D_2, the equilibrium price is P_2, which differs little from P_1.

Much more important for trends in real mineral commodity prices are shifts in the long-run supply curve. Mineral depletion, rising wage rates, and other factors that push costs up shift the curve upward, while new technologies that reduce the costs of finding and processing mineral commodities shift the curve downward. If over time the cost-reducing effects of new technology are able to offset the cost-increasing effects of depletion and other factors, then real prices will trend downward, whether demand is growing slowly or rapidly.

Figure 4.3b illustrates this situation. It assumes that, thanks to new technology, the long-run supply curve moves down from S_0 to S_1. Now, the long-run equilibrium price is P'_1 if demand grows slowly and the demand curve shifts only to D_1 and is P'_2 if demand grows rapidly and the demand curve shifts to D_2. Again, these two prices are nearly the same, suggesting that the pace of demand growth

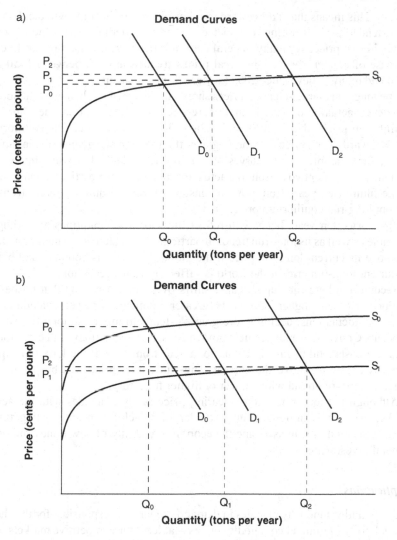

Figure 4.3 Long-run price trends: (a) the influence of demand; (b) the influence of demand and supply.

does not greatly affect the long-run equilibrium price. More interesting is the fact that both of these prices are considerably below the prices (P_1 and P_2 in Figure 4.3a) associated with the supply curve S_0, and which would have prevailed had the supply curve not shifted downward.

This explains why the long-run trend in real prices has at various times for many mineral commodities sloped downward. This was the case, as we discussed earlier, for all the commodities shown in Figure 4.1 over the years 1970–2003. Of course, the long-run supply curve can shift upward and does so when new technology is unable to offset the cost-increasing effects of depletion and other factors. This means that a downward trend in real prices is not inevitable. Indeed, for extended periods during the nineteenth and twentieth centuries the long-run trends in real prices for many mineral commodities were stagnant or positive. In the case of copper, for example, real prices trended upward between 1950 and 1970 and many believe since 2003 they are doing so again.

The long-run market-clearing prices shown in Figures 4.3a–b, it is worth noting, are never actually observed. In this respect they differ from the short-run equilibrium prices shown in Figures 4.2a–b. Rather, they reflect the equilibrium levels toward which prices are moving over the long run at a given point in time—today, for example, if the curves shown reflect today's long-run supply and demand curves. For two reasons the actual market price for a particular commodity in the future (ten years from now, for instance) may be quite different from its current long-run equilibrium price.

First, price ten years hence will reflect the short-run conditions prevailing at that time as well as its long-run trend. In particular, the actual price then will likely be above its current long-run price if the global economy is booming and below its current long-run price if the world is suffering from a recession.

Second, the long-run supply curve shifts over time. Anything that pushes up production costs—higher labor costs, greater equipment and construction costs, currency fluctuations, declining ore grades—tends to move the long-run supply curve up. Conversely, management innovations, new technology, and other forces reducing costs tend to shift it down. So, a year from now, if the long-run supply curve is below or above the current long-run curve, price will be pursuing a different long-run trend with a lower or higher trajectory.

Although the long-run market-clearing price rarely coincides with the actual market price, it is important. In particular, it is helpful in anticipating future price trends and thus in assessing the economic viability of new mines and other mineral investments.

Implications

We now understand why changes in demand are largely responsible for the short-run volatility of mineral commodity prices—at least for competitive markets with limited unused capacity—while changes in supply largely determine the long-run trends. We turn next to some of the implications of these findings.

Firm Diversification

As noted earlier, intense waves of mergers and acquisitions have swept over the mineral sector from time to time during the past half-century. Acquiring firms often claim that one of the benefits of merging with companies that produce other mineral commodities is a reduction in market risk. Given the volatility of prices, the profits of firms that produce just one or two mineral materials will also be volatile. Mergers can reduce a firm's dependence on one or a few mineral commodities and, so it is argued, reduce its exposure to market risk.

This is true, however, only if mineral commodity prices do not move closely together. Given that short-run price fluctuations are largely driven by shifts in the demand curve, the result of macroeconomic swings in the business cycle, we should expect that mineral commodity prices are quite synchronized. Demand and prices are up when the global economy is booming and down when the world is suffering from recession. As Table 4.2 shows, the empirical evidence indicates that this is the case. Annual changes in average real prices among the major mineral commodities are all positively correlated. For many, the correlation coefficients are 0.60 or higher. As a result, the benefits of lower market risk associated with diversification can easily be exaggerated when diversification simply means the production of other mineral commodities.

Forecasting Prices

A second implication is that short-run and long-run price forecasts, which producing companies and countries need for planning and other purposes, require quite different sets of information, skills, and experience. To forecast mineral commodity prices next year, expertise in macroeconomics is essential, and in particular an ability to predict the growth of the construction, capital equipment, transportation, and consumer durable sectors—or alternatively of industrial production—around the world. Of course, additions to capacity and closures also need to be taken into account, though this information is normally readily available from company news releases, the trade press, and other sources, given the lead times involved with such developments.

Good forecasts of mineral commodity prices 10–20 years in the future, on the other hand, depend much more on in-depth knowledge of current average total costs (that is, capital and operating costs) of the marginal mines and producers, and an understanding of how these costs are likely to change over time. This requires an ability to spot the important new technological developments on the horizon and to assess how they will alter the costs of finding, mining, and processing minerals over the years to come. Here, much more engineering, science, and microeconomics are needed, and much less macroeconomics, than for short-term forecasts.

Promoting and Protecting Demand

Mining firms, with a few exceptions such as gemstone producers, spend very little on developing and protecting demand, many times less than they spend on exploration or on reducing production costs. This is in part because the benefits from creating new demand are enjoyed by all producers, those that support such collaborative efforts and those that do not. This fosters the free-rider syndrome, a natural tendency for individual firms to let others shoulder what should be a collective responsibility. This is particularly a problem in competitive industries with many producers. As a result, some argue that the mineral industries are missing an important opportunity and should increase their efforts to promote new demand.

Table 4.2 Correlation coefficients for annual changes in average real prices between phosphate rock, potassium chloride, aluminum, copper, lead, tin, nickel, zinc, gold, platinum, silver and iron ore 1970–2012

	Phosphate rock	Potassium chloride	Aluminum	Copper	Lead	Tin	Nickel	Zinc	Gold	Platinum	Silver	Iron ore
Phosphate rock	1.00											
Potassium chloride	0.79	1.00										
Aluminum	0.14	0.06	1.00									
Copper	0.56	0.64	0.45	1.00								
Lead	0.58	0.63	0.51	0.86	1.00							
Tin	0.42	0.26	0.41	0.37	0.61	1.00						
Nickel	0.36	0.45	0.58	0.79	0.75	0.17	1.00					
Zinc	0.29	0.28	0.36	0.73	0.60	0.17	0.72	1.00				
Gold	0.57	0.73	0.11	0.62	0.62	0.42	0.46	0.27	1.00			
Platinum	0.59	0.75	0.36	0.77	0.74	0.27	0.74	0.43	0.84	1.00		
Silver	0.52	0.54	0.33	0.56	0.67	0.74	0.35	0.24	0.85	0.68	1.00	
Iron ore	0.73	0.77	0.27	0.82	0.80	0.38	0.74	0.46	0.79	0.89	0.64	1.00

Source: World Bank (annual).

Others are concerned that high prices will hurt demand growth by encouraging the substitution of plastics and other alternative materials for mineral commodities and by stimulating material-saving new technologies. Moreover, as Chapter 2 points out, the effects of such developments are unlikely to be reversible, at least completely so, when prices decline.

While these concerns merit attention, their implications for mineral commodity prices and output in the long run are quite different. Whether demand grows modestly or briskly has little influence on the long-run price, but a substantial impact on output and the size of markets. As a result, producers interested solely in maximizing profits from existing operations have less to fear from falling demand (and thus less to gain from collective efforts to promote demand) than those who desire to grow over the long run and need an expanding market to do so.

China and Long-Run Commodity Price Trends

Many mineral economists and market analysts believe that the declining long-run trend in real prices for many mineral commodities since 1970 (see Figure 4.1) has come to an end and is now rising. The evidence they advance in support of this conclusion includes the sharp rise in real prices since 2003, rising demand in China and other developing countries, and higher production costs.

The recent jump in prices, however, might simply be the result of short-run circumstances. Mineral commodity prices, as we have seen, are known for their volatility, and so higher prices for a few years need not necessarily portend a change in the long-run trend. They could instead reflect the failure of the mineral industries to add sufficient new capacity in a timely manner over the past decade due in part to their failure to anticipate the magnitude of the surge in demand from China and other developing countries. Of course, rising demand in China and elsewhere may continue for some time. But, as highlighted earlier, growth in demand, even if persistent, has little impact on long-run prices when the long-run supply curve is relatively flat over the likely range of future prices.

What matter are changes in production costs, as it is these changes that shift the long-run supply curve up and down. This means that those who contend that the secular trend in real prices is now upward need to explain why they believe that the cost-reducing effects of new technology, which for decades have offset the cost-increasing effects of mineral depletion and the need to exploit poorer quality mineral deposits, are no longer able to do so.

Investors and Speculators

This section explores the influence of investors and speculators on commodity markets and prices, which up to this point we have largely ignored. The role of these market participants is controversial, still shrouded in mystery and far from completely understood. Many believe that investors and speculators accentuate price volatility over the short run and alter long-run trends in real prices. Others disagree.

Investors and Price Volatility

Corporate officials, market analysts, consultants, and others closely associated with commodity markets widely believe that investors accentuate price volatility. When prices are rising, they enter the market and drive the price up further. Then when the price starts to fall, they sell their holdings, causing the price to drop even further.

Mineral economists and economists more generally, however, often take issue with this view. They contend that if investors are actually destabilizing prices they must be buying when prices are relatively high and selling when they are relatively low. In this case, they are losing money and should be driven out of the market. On the other hand, if they are making money and hence have an incentive to remain active as investors, the price at which they are selling must be higher than that at which they are buying. In this case, they are reducing price volatility.

Efforts to assess empirically the impact of investors on price volatility are hampered by serious deficiencies in the available data. As a result, both sides in this debate can continue to maintain that they are right and the other wrong.

Investors and Secular Price Trends

A somewhat similar though different debate surrounds the influence of investors on long-run commodity price trends. Some analysts maintain that fundamentals—that is, the supply and demand of producers, consumers, traders, and governments—govern secular price trends. Others contend that the surge in investor demand, particularly in commodity hedge funds, is largely behind the substantial rise in their prices since 2003.

Figures 4.4a–b, though both reflect short-run supply and demand curves, are helpful in thinking about this controversy. The first of these figures shows that the market price (P_1)—the spot price, for example—for a mineral commodity depends on its market supply curve (SS) and its total market demand curve (TD_1). This latter, in turn, depends on the consumer demand curve (CD_1) and the investor demand curve (ID_1).

At the market price P_1, consumers will demand the quantity $0CQ_1$ and investors the quantity $0IQ_1$ (which they add to their existing inventories). Total demand is the sum of the two or $0TQ_1$. Investor demand, it is worth noting, can be negative. In Figure 4.4a, for example, as price rises above P_1^*, investors on balance start depleting their inventories and adding to supply.

As drawn, both the consumer and investor demand curves have negative slopes. This seems reasonable for the consumer demand curve, but the investor demand curve may not have a negative slope over its entire range. This is because investor demand, among other things, depends on both the current price and the price that investors expect to prevail in the future. When the current price rises, this may cause expectations about prices in the future to rise as well. If so, a rise in the current price may cause investor demand to increase. However, as prices continue to rise, at some point, the slope of the investor demand curve will become negative.

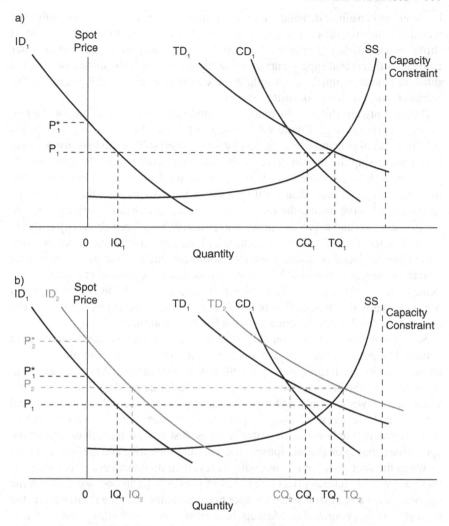

Figure 4.4 Hypothetical spot supply and demand curves for a commodity illustrating rising prices and stocks due to an increase in investor demand. (a) Period 1: initial market equilibrium given the producer supply curve (SS), consumer demand curve (CD_1), investor demand curve (ID_1), and total demand curve (TD_1). (b) Period 2: market equilibrium after a rise in investor demand shifts the investor demand curve (from ID_1 to ID_2) and total demand curve (from TD_1 to TD_2).

This is because at higher and higher current prices it becomes increasingly difficult for investors to maintain the expectation that prices in the future will continue to rise. Thus, at some price the slope of the investor demand curve will become negative, as shown in Figures 4.4a–b.

Commodity prices can rise over time as a result of an inward shift in the market supply curve, an outward shift in the total demand curve (due to an outward shift in either the consumer demand curve or the investor demand curve), or both. The recent rise in mineral commodity prices, we know, is not the result of inward shifts in the market supply curve. Growth in production capacity and other evidence suggest that supply curves, while they may not have shifted outward in some instances as much as we might have expected given higher commodity prices and profits, have not shifted inward.

Rather, outward shifts in the market demand curve are responsible for the rise in prices in recent years. Figure 4.4b illustrates this situation. It shows an outward shift in the total demand curve, caused by an outward shift in the investor demand curve, increasing the market price. A similar figure could show the same rise in price caused by a comparable shift in the consumer demand curve rather than the investor demand curve. What we do not know is to what extent the shifts in the total demand curve behind the recent rise in prices have been driven by shifts in the investor demand curve and to what extent they have been driven by shifts in the consumer demand curve. The empirical studies undertaken to resolve this issue have produced conflicting results—some concluding that greater consumer demand is largely responsible for higher commodity prices and others that investor demand is mostly responsible. Until the deficiencies in the available data and other problems that plague this research are resolved, the controversy over the role of investor demand on price trends is likely to continue.

So far we have focused on the influence of investor demand on mineral commodity prices when both are on the same market—for example, the spot market, the 30-day futures market, the 90-day futures market. Another intriguing issue concerns the influence on spot prices of investor demand on futures markets. Since most investors have traditionally not wanted to take delivery of physical stocks, they have invested largely by purchasing futures contracts and then selling these contracts at or before their maturity dates. Just how such purchases affect the spot price depends on the relationship between the spot price and the futures price.

When the spot price for a commodity is less than its futures price (or prices, as there are various futures prices), the market is said to be in *contango*. When the opposite is the case, that is when the spot price is greater than the futures price, the market is in *backwardation*. Markets in contango can be further separated into those in strong contango (where the spread between the futures and spot price is sufficient to cover the costs of buying and holding physical stocks) and those in weak contango (where the spread is not sufficient to cover the costs of buying and storing the commodity).

When markets are in strong contango, any increase in investor demand that drives up the price of a commodity in its futures market should have an immediate and comparable effect on its spot price. This is because a rise in the futures prices will encourage investors to buy on the spot market and simultaneously sell short in the futures market. They then hold their acquired inventories to cover their commitment to deliver the commodity in the future. These transactions will cause the spot price to rise and the futures price to fall until the spread between the two no longer encourages further inter-temporal arbitrage. Given this situation, spot and futures prices should be closely correlated, and so investor demand or anything else that causes the futures price to rise should similarly impact the spot price.

However, when markets are in weak contango or backwardation, investors do not have an incentive to buy in the spot market and sell short in the futures market when investor demand increases price in the futures market. As a result, prices on the spot market are likely to be less correlated with movements in the futures price. Under these market conditions investor demand in the futures market may have little influence on the spot price. In short, the answer to the question of whether investors in futures markets affect the spot price may depend on whether or not the spot and futures markets at the time are in strong contango.[4]

Highlights

Metals and non-metallics are bought and sold in a variety of ways. Prices may be set by commodity exchanges, in over-the-counter (OTC) markets, by producers, or by negotiation between buyers and sellers. Market participants include both those interested in physical stocks (producers, consumers, traders, and governments) and those interested in commodities simply as paper assets (investors and speculators). Many mineral commodity markets, though not all, are competitive. In competitive markets no buyer or seller possesses market power, which requires both the ability to influence the market price and the incentive to do so.

Public policy generally favors competitive markets because they seem more equitable and they promote both production efficiency and allocative efficiency. They may also be better at promoting dynamic efficiency by stimulating the creation and diffusion of new technology, though this is less clear.

The past half-century has witnessed many new developments in metal and non-metallic markets—the rise and decline of state mining enterprises; the entrance and exit of large petroleum companies; the shift toward open market economies in China, the former Soviet Union, and Eastern Europe; new technologies greatly reducing the costs of shipping bulk commodities; rapid global growth in commodity consumption; and periodic waves of mergers and acquisitions. Overall these events have fostered competition and have converted many commodity markets where once producers or buyers possessed market power into competitive markets.

Mineral commodity prices are known for their short-run volatility, which arises because the demand varies greatly over the business cycle. When GDP is up, the sectors that consume most metal and non-metallic output—construction, capital equipment, consumer durables, and transportation—are up even more. So demand in the short run is very responsive or elastic with respect to income. However, it is not very elastic with respect to price. Nor, once full capacity is approached, is supply in the short run very responsive to changes in price. This means that when output is near the capacity constraint, both the supply and demand curves are very steep. As a result, large changes in the market price are needed to keep them in balance over the business cycle as the demand curve shifts to and fro. Producer prices can dampen the fluctuations in price, but market volatility then simply manifests itself in greater sales volatility and in actual physical shortages during the boom of the cycle.

While short-run fluctuations in mineral commodity prices are largely the result of shifts in the demand curve, long-run trends in real prices are primarily the result of shifts in the supply curve. This is because the long-run supply curve is not constrained by capacity and so does not turn vertical at some output. Rather, it tends to rise with output initially, thanks to a limited number of high-quality, low-cost deposits, and then to flatten out, as marginal deposits are much more common and numerous. As a result, whether the demand for mineral commodities in China and India grows slowly or rapidly over the next couple of decades will not have a great impact on the long-run price (though it will influence output). The trend in long-run prices will, instead, depend much more on whether new technology can offset effects of depletion and other factors tending to push production costs higher.

Finally, the role of investors and speculators on mineral commodity prices remains something of an enigma. Some argue that these market participants accentuate the volatility of mineral commodity prices and distort their long-run trends. However, others are dubious, and the jury is still out on this debate. Its resolution awaits both better data and additional research.

Notes

1   This section draws heavily from Humphreys (2011).
2   The non-metallics, perhaps due to their nature, were less of a target for nationalization.
3   Figure 4.1 also shows the relationship between price and time estimated by simple regression analysis. The coefficient on the time (t) variable indicates the annual reduction in the long-run price trend over the 1970–2003 period. For example, in the case of aluminum the trend in real prices declined by 38.02 dollars per year.

   Also shown is the coefficient of determination (R^2). It indicates how much of the year-to-year change in price over the 1970–2003 period is associated with the long-run downward trend in prices. In the case of aluminum, the estimate for the coefficient of determination is 0.28, suggesting that the long-run downward trend in real prices can explain up to 28 percent of the yearly variation in the aluminum price. The rest arises as a result of changes in the business cycle and other factors causing short-run fluctuations in the aluminum price.
4   Recent research (Gulley and Tilton, 2014) estimates the correlations between LME spot and futures copper prices during periods of strong contango, weak contango, and backwardation. While it finds that the correlations are highest during periods of strong contango, they are also quite high during periods of weak contango and backwardation as well.

References

Gulley, A. and Tilton, J.E., 2014. The relationship between spot and futures prices: an empirical analysis. *Resources Policy*, 41, 109–112.

Humphreys, D., 2011. Pricing and trading in metals and minerals, in Darling, P., ed., *SME Mining Engineering Handbook*, third edition, Society for Mining, Metallurgy, and Exploration, Littleton, CO.

Lundgren, N.G., 1996. Bulk trade and maritime transport costs. *Resources Policy*, 22, 5–32.

Scherer, F.M., 1980. *Industrial Market Structure and Economic Performance*, second edition, Rand McNally, Chicago.

Tilton, J.E., 1971. *International Diffusion of Technology: The Case of Semiconductors*, Brookings Institution, Washington, DC.

UNCTAD (United Nations Conference on Trade and Development), annual. UNCTADSTAT. Available at: http://unctadstat.unctad.org/ReportFolders/reportFold ers.aspx

World Bank, annual. *World DataBank: Global Economic Monitor (GEM) Commodities*. Available at: http://databank.worldbank.org/data/reports.aspx?source=global-econom ic-monitor-(gem)-commodities

Part II
Public Policy

5 Public Policy, Rents, and Taxation

Earlier chapters have focused on the nature of market forces—particularly supply and demand—that explain the behavior of mineral commodity markets. Part II of this book explores the role of public policy in the mineral sector. This chapter begins this discussion by examining the rationale for mineral policy. Where and when should governments intervene in commodity markets and attempt to alter behavior from what the private marketplace would produce? It then explores the nature of mineral rents and user costs, which leads finally to a discussion of mineral taxation.

Subsequent chapters in Part II address other mineral policy issues: international trade and comparative advantage in mining; market power and competition; mining and economic development; depletion and mineral shortages; and finally environmental concerns and sustainability.

Rationale for Government Policy

A wide spectrum of views exists on the proper role of government in the economy in general and the mineral sector in particular. At one extreme are anarchists, who see no useful role for government in any domain. At the other extreme are the pure socialists, who see no role for the private sector. They would place all the means of production, including mining and mineral processing, into the hands of the state.

Most mineral economists reject both of these extremes. They favor leaving mining and mineral processing in the hands of private companies, in large part because history suggests these entities are often more efficient than their state-owned counterparts. But they also recognize that the government needs to provide the laws, regulations, and other rules of behavior as well as the police, courts, and other institutions to enforce them.

Moreover, the private sector without government intervention suffers from various *market failures*—outcomes that fail to optimize the welfare of society. For example, mining companies in the absence of regulations pollute the land, air, and water more than is desirable. Even if private companies would like to be good citizens and restrict their pollution, doing so raises their production costs. This reduces their competitiveness and may even force them out of business. Market failures may also arise if private companies have a shorter time horizon or greater risk aversion than the public at large. This will cause them to skew their investments too far toward low-risk, quick payoff projects.

History suggests as well that central banks and governments can play a useful role in mitigating fluctuations in the business cycle, reducing inflation during economic booms and unemployment during recessions. In addition, public policies can alleviate the high concentration of wealth and income that the private marketplace on its own often creates. Many mineral economists also believe that governments should pursue policies to promote competitive commodity markets, free of monopoly or market power; to ensure that the domestic mineral sector promotes sustainability and economic development; and to alleviate shortages arising from mineral depletion and other causes.

Finally, most mineral economists and other economists believe that governments have an obligation to tax the mineral sector, as they tax other sectors, in order to obtain the financial resources needed to provide for the national defense, public safety, education, and other public goods. If left to the private sector, these services and goods would either be greatly underfunded or not funded at all. In addition, the taxation of mineral production is often recommended for equity reasons, as it allows governments to capture some of the country's resource wealth and mineral rents for society as a whole.

For all these reasons, government intervention in mineral markets may be desirable. However, it is important to stress that public policies can suffer from what are called *government failures*. Poorly designed and executed policies may entail more costs than benefits for society. So we cannot automatically presume that a market failure justifies government intervention. For this reason, economists and policy analysts do not always agree on where and when government intervention is justified. Those with faith in the efficacy of public policy tend to favor more government intervention than those who lack this faith.

In assessing the performance of public policy, it is helpful to consider the goals—or what economists call *the objective function*—of the government and public policy. In the case of private firms, we normally assume the objective function is to maximize profits. Or, in an inter-temporal setting where, for example, higher profits this year may lead to lower profits in future years, we assume their goal is to maximize the net present value (NPV) of current and future profits (see Box 4.1).

Of course, this need not be the case. Many years ago one of us (John Tilton) visited a small gold mine in northern Quebec with some graduate students. The mine manager showed us data on his production for a number of years. Somewhat surprisingly output had a clear tendency to fall when the price of gold rose and to rise when the price of gold fell. When asked why he did not maximize the mine's NPV by producing more when prices were high, he replied the mine was the major employer and source of income for the local community. Since its reserves were limited, by mining lower grade ore when prices were high he could both meet the profit expectations of the parent company based in New York and extend the life of the mine to the benefit of the local community. This manager's objective function was to maximize the life of his mine subject to a profit constraint that kept his superiors in New York happy. Over the past 50 years economists have identified many other possible objective functions that private firms may pursue. However, we still assume that the objective function of private firms in most instances is profit or NPV maximization.

For governments, however, this objective function is clearly not applicable. Governments are not in the business of earning profits, but rather are supposed to be serving the interests of their citizens. This would suggest that the objective function for public policy is to maximize society's welfare. Defining just what is the social welfare, of course, is tricky. In a democratic society, presumably social welfare is defined through the interplay of various interests in the political process along with the laws and other outcomes that this process produces.

Of course, there are caveats. Some would argue that even in democratic societies the political process is so imperfect that it mostly serves special interests rather than society as a whole. Moreover, where there is little or no democracy, public policies may strongly favor one ethnic or tribal group over others or promote a state religion, even though this reduces the social weal. Few would argue, for example, that the governments of North Korea, Iran, or Zimbabwe are primarily focused on the welfare of all their citizens. Nevertheless, we normally assume that the objective function of government policy is to maximize the social welfare (as defined by a well-functioning democratic process), just as we assume despite the exceptions that the objective function of private firms is to maximize NPV.

Mineral Rents[1]

Earlier we noted that one reason for government intervention is to ensure that society as a whole shares some of the wealth or profits arising from the exploitation of domestic mineral resources. This is usually accomplished by taxing mineral production. To assess public policies in this area requires an understanding of how mineral wealth is created and in turn the nature of mineral rents.

Economists define *economic rent* as a surplus of income that can be taken away from firms and individuals without altering their economic behavior. For example, a soccer player who is also a good mineral economist may have the option of playing soccer at a salary of 1,000,000 dollars per year or of working as a mineral economist at a salary of 200,000 dollars per year. If he is motivated solely by salary and willing to take whichever position pays the most, then he will play soccer at a salary of 1,000,000 dollars. Of this amount, 800,000 dollars is economic rent, since the government could tax this amount away without altering his economic behavior.[2]

Similarly, the economic rent that a firm realizes from a mining project—often called *Ricardian rent* after the nineteenth-century British economist David Ricardo, who analyzed rent particularly for agricultural land of different quality— is the revenue it earns above that amount just required to bring it into operation or, if the venture is already in operation, to keep it in production. For an operating mine, for example, the Ricardian rent that the owner earns per unit of output is the difference between the price it receives and its average variable (or out-of-pocket) costs of production. As long as the market price exceeds its average variable costs, the firm has an incentive to keep the mine in operation even though it may not be recovering its capital costs (as capital costs are sunk and have to be paid whether or not the mine is closed).

The Ricardian rents earned by firms vary from one mine to another, as Figure 5.1a illustrates. This figure reproduces the comparative cost curve shown in Figure 3.9 and adds the market price (P_m). The comparative cost curve, it will be remembered, shows the cash costs or average variable costs of operating mines, from the mine with the lowest to the mine with the highest costs.

Assuming mines remain in operation as long as—and only as long as—they are recovering their average variable costs, then at the market price P_m Mines A through G will be in operation. Mine H with average variable costs above the market price will be shut down. The rent that each mine realizes is the difference between the market price (P_m) and its average variable costs. So the marginal mine—Mine G—earns no rent. Mine A, on the other hand, with the lowest average variable costs ($0C_a$) enjoys the greatest rent per unit of output (C_aP_m). The total rent this mine earns for its owner equals its rent per unit of output (C_aP_m) times its output ($0Q_a$). Mine B enjoys the second largest rent per unit of output, Mine C the third, and so on. Although these mines all realize a smaller rent per unit of output than Mine A, their total rent may exceed that of Mine A if their output is greater (as is the case for Mine B).

The Ricardian rent earned by individual mines arises for various reasons. First, there are the *pure rents*. These occur because some mineral deposits are of higher quality and so cheaper to exploit than others thanks to their ore grade, size, depth of overburden, ease of processing, location, or other attributes.[3]

Second, there are *monopoly rents*. These occur when mining firms possess market power, which allows them to earn excess profits even over the long run without inducing new entry.

Third, there are *rents due to short-run supply constraints*. Even in competitive industries where entry is possible, price may rise above the long-run competitive level when demand is strong due to the time it takes to build new capacity. The excess profits this generates will encourage new capacity and eventually cause the price to fall, but in the interim existing mines realize rents due to short-run supply constraints. These rents exist only over the short run.

Fourth, there are *quasi rents*. These, too, exist only over the short run. They arise because capital investments in mines, once made, are a sunk or fixed cost. So, as noted before, mines will stay in operation as long as the market price is above their average variable costs, even though they may not be recovering all of their capital costs. New mines, however, will not come on-stream unless the market price is sufficiently high to permit the recovery of both their variable and fixed costs. This limits supply so that existing mines in the short run normally earn a quasi rent that approximates the average fixed costs of mine production.

Finally, there are *rents due to public policy*. Some mines enjoy low average variable costs because their taxes are low or because they are subjected to less onerous government regulations than other mines. This source of rent, it is

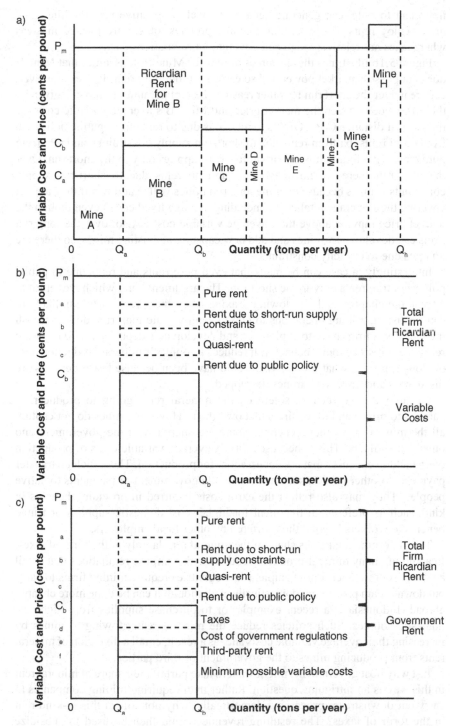

Figure 5.1 (a) Ricardian rent for Mine B; (b) pure rent, rent due to short-run supply constraints, quasi-rent, and rent due to public policy for Mine B; (c) rent from Mine B captured by government and third parties.

important to note, can generate negative as well as positive rents. While some mines enjoy rents due to favorable public policies, others are located in areas where taxes are relatively high and government regulations costly.

Figure 5.1b illustrates these sources of rent for Mine B. It assumes that Mine B does not possess market power and so earns no monopoly rent. It does, however, earn rent over the short run for other reasons. Its rent per unit of output is given by the difference between the market price and Mine B's average variable costs, or the vertical distance C_bP_m. Of this total, C_bc is due to rents from public policy, cb from quasi rents, ba from rents due to short-run supply constraints, and aP_m from pure rents. This figure, it is important to stress, applies only to the short run. Over the long run there are no quasi rents and no rents due to short-run supply constraints. This is because over the long run mines will shut down if they are not covering their average total costs (including average fixed costs) even though the market price may be above their average variable costs. Also, over the long run firms can develop new mines and add new capacity at existing mines, so there are no rents due to capacity constraints.

Interestingly, a case can be made that even pure rents and rents due to public policy are true rents only in the short run. The argument here, which we return to later in the chapter, is the following: Taxing away these rents, while unlikely to cause existing mining operations to close even over the long run, does diminish the incentives firms have to explore for and develop new deposits. So taxing these rents does alter economic behavior. It reduces a country's mineral production over the long run from what it otherwise would have been because fewer deposits are discovered and fewer new mines developed.

So far, we have focused solely on the mineral rents going to producers—namely, to mines and to the firms that own them. However, mines do not capture all the mineral rents they generate. Some are shared with the government and other stakeholders. This is because a firm's average variable costs of production often include expenses that are not essential for production. These include transfer payments to other parties, such as taxes to the government or payments to native peoples. They may also reflect the extra costs incurred in providing benefits in kind, such as preferential treatment for higher-cost domestic supplies or fringe benefits to workers beyond those offered by other local employers.

The government takes its share of the mineral rent largely in the form of taxes. In addition, many mineral-producing countries also impose regulations that entail a type of rent transfer. For example, some countries require mining firms to carry out downstream processing domestically even though it can be done more cheaply abroad (Indonesia is a recent example) or to purchase supplies from high-cost domestic sources. Such policies reduce the mineral rents flowing to mines by increasing their average variable costs and, as a result, entail a diversion of mineral rents from producing mines to the government or third parties.

Just why host governments might want to take part of their share of mineral rent in this way is an intriguing question. Rather than requiring mining companies to carry out downstream processing domestically, why not collect this in-kind rent in the form of taxes? The resulting revenues could then be used to subsidize more labor-intensive industries if employment is a concern, more skill-intensive industries if the development of a skilled labor force is a priority, less polluting industries if the quality of the environment is an issue, or more R&D-intensive industries if enhancing domestic research and development capabilities is a goal. The answer is not clear. It may be that government officials are simply unaware that imposing unnecessary cost-increasing regulations reduces the share of total rents that the government can collect in the form of taxes.[4]

Other stakeholders may also share in the rents generated by mining. These include local communities, organized labor, environmental organizations, and indigenous peoples. Indeed, any other group that can impede the development and operation of a mining project possesses the power to obtain some of the rents that it generates.

During the 1950s and 1960s, for example, when the US copper industry was highly competitive and dominated world production, labor unions in the country successfully bargained for wages that exceeded by wide margins those prevailing in comparable jobs elsewhere in the economy. These high wages reflected a transfer of some of the rents from the mine owners to workers. When the industry fell upon hard times in the 1980s, workers accepted a sharp cut in wages to keep their companies from bankruptcy and thus to preserve their jobs. These jobs were worth preserving because their wages were competitive even after the reductions.

Similarly, when local communities, indigenous peoples, environmental groups, and others win cost-increasing concessions from mining companies, they are in effect capturing some of the rent that mining creates.

Figure 5.1c illustrates the division of mineral rents for Mine B among various parties. The total rent per unit of output generated by this mine is given by the difference between the market price (P_m) and the mine's minimum possible variable costs (f). Of this total, the mine and its owner capture C_bP_m, the government captures eC_b in taxes and in-kind transfers due to regulations, and other stakeholders (labor, environmental organizations, local communities, indigenous peoples) capture fe.

Hotelling Rent and User Costs

Mineral commodities are non-renewable, at least on any time scale of relevance to the human race. In a seminal article published in the early 1930s (Hotelling, 1931), the American economist Harold Hotelling argues that a mine with an exhaustible stock of reserves will behave differently from other firms. Normally we expect that a competitive firm will continue to increase output until the cost of producing the next unit just equals the market price it receives for that unit. However, in addition to its production costs a mine must take into account the opportunity cost associated with producing one more unit of output during the current period. This cost arises because reserves used today are not available in the future. It equals the NPV of the loss in future profits due to producing one more unit of output today. These losses arise either because there are fewer reserves to exploit in the future, as Hotelling assumes, or because remaining reserves are of lower quality and hence more costly to mine and process.

Mineral economists and economists in general refer to this opportunity cost as *user costs*, *Hotelling rent*, and *scarcity rent*. These three terms all mean the same thing. We prefer and use the term user costs because this cost is in fact not a rent. If mines cannot recover their production costs plus user costs, they have an incentive to alter their behavior by producing less now and more in the future.

Hotelling focuses on an individual mine. Others have since applied his analysis to the mining sector as a whole. Figure 5.2 shows how this extends our earlier analysis of mineral rents. Now the marginal mine (Mine G) is prepared to operate only if the market price is at the level P_m', which is sufficiently high to cover both its cash costs of production and its user costs. At a price lower than P_m', the mine is better off keeping its reserves in the ground for use in the future.

We can think of user costs as the value of reserves of marginal quality (i.e., the quality of Mine G's reserves) in the ground. The reserves of mines with lower costs (Mines A through F), of course, are more valuable because they generate Ricardian rents thanks to their higher quality. Moreover, under ideal conditions exploration should continue until the expected cost of finding new marginal reserves just equals the value of such reserves in the ground. This suggests that user cost can be approximated both by the value of marginal reserves in the ground and by their expected discovery cost.

Mineral reserves differ from renewable resources, such as agricultural land, in that they, in theory at least, have an additional value equal to their user costs thanks to their exhaustible nature. Since mineral reserves in many countries belong to the state, many contend that mining companies exploiting these subsoil assets should compensate the government by paying a tax that captures the user costs. We explore this idea more in the next section on mineral taxation.

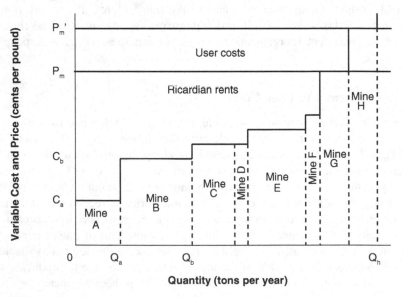

Figure 5.2 User costs for Mines A to H.

Mineral Taxation

Governments around the world impose a wide variety of taxes on their mineral producers. There are, for example, taxes on the imports of machinery and equipment needed to develop new mines. There are fees for converting foreign exchange and for transferring dividends abroad. Here, however, we focus on the three most important types of mineral taxes—royalties, corporate income taxes, and excess profit taxes—and on two particular issues or questions. First, what is the proper mix of taxes? Should mineral-producing countries rely solely on corporate income taxes (as they do with most other industries) or should they impose a mix of taxes? Second, what is the optimum level of taxation?

The Mix of Taxes

Before delving into these two issues, we should be clear on how these three taxes—royalties, corporate income taxes, and excess profit taxes—differ. In practice, there is considerable confusion. One recent study (Otto *et al.*, 2006), for example, concluded that a royalty was a royalty when the government called it a royalty. Chile, for example, imposes what it calls a royalty on mining companies that actually depends significantly on the production costs and profits.

Here we define royalties as taxes on a company's mine production measured either in physical terms, such as tons, or monetary terms, such as the dollar value of its output. So royalties are determined solely by output and perhaps market prices, but not by company profits or losses.

Corporate income taxes, in contrast, depend on a company's net income or profits. This tax may be a fixed percentage of a company's profits or it may be progressive, with its rate increasing as profits rise. In the United States, for example, the marginal corporate income tax rate rises from 15 to 35 percent as profits increase. While royalties and excess profits taxes are imposed for the most part just on companies producing mineral commodities, most countries impose a corporate income tax on firms operating in other economic sectors as well.

Excess profit taxes—also known as resource rent taxes—are designed to capture a portion of unusually large profits or rents, which mining companies may earn as a result of a boom in commodity prices or simply from possessing unusually rich and hence low-cost deposits. These taxes may be triggered when the internal rate of return or NPV of projects exceeds a particular threshold or when commodity prices rise above prescribed levels.

This brief summary of the three principal types of mineral taxes raises the question: Why does the mineral sector have three common types of taxes, rather than just corporate income tax like most other economic sectors? A good part of the answer lies with: (a) public perceptions regarding mineral wealth; (b) the risk and great dispersion in returns to mineral projects; and (c) the high political visibility of mining, especially in developing countries where this sector contributes greatly to overall economic performance.

For reasons discussed in the previous chapters, mineral projects produce a wide spectrum of returns. Some are extremely lucrative. Others lose hundreds of millions of dollars. The corporate income tax divides this risk between companies and countries in an acceptable manner much of the time. Troubling problems arise, however, from the perspective of the public and hence the government at both extremes—that is, when profits are very high or very low.

When mining ventures lose money, the corporate income tax produces no government revenues. Nevertheless, these projects can consume what many consider valuable domestic mineral resources. They may also require government resources to oversee and regulate. At the other extreme, when projects do very well and earn great profits, government revenues from a corporate income tax are high. However, companies are likely to be doing even better, earning extraordinary returns on their investments. During its first year of operation, for example, the Panguna mine on Bougainville Island in Papua New Guinea is reported to have earned about 40 percent of its total investment (Mikesell, 1975). Such high rates of return create problems for producing countries. The general public and government officials feel the country is not receiving a sufficient share of the rents arising from their geological heritage.

Royalties ensure that producing countries receive some tax revenue even when mines are earning little or no profits. Excess-profit taxes allow host governments to increase their share of rents when producers are earning unusually high profits. As a result, both types of taxes are often found in mineral-producing countries.

Private companies understandably are not terribly enthusiastic about these special taxes on mining, and in particular object to excess-profit taxes. These taxes may, however, have benefits for companies as well as countries. In particular, they help ensure the viability of the tax regime and other agreements that mining companies have with host governments. Projects that earn a 40 percent rate of return for foreign mining companies are politically viable in very few countries. Papua New Guinea, for example, renegotiated the Bougainville agreement after the mine's first year of operation.

As the next section highlights, there is an optimum level of taxation that producing countries can impose on their mineral companies. This means that governments have to make a choice on how much of their expected tax revenues they want to come from royalties, corporate income taxes, and excess profit taxes. The proper balance among these three types of taxes varies from country to country. It depends on just how much of the risk associated with mining ventures a government wants to assume and on how willing it is to allow private companies to earn sufficient profits to compensate for the share of risks that they assume.

The Optimal Level of Taxation

We start the discussion of this issue by examining—and questioning— three common arguments frequently advanced in support of greater taxes on mineral ventures.

1   Some mineral deposits, largely as a result of the geological forces that created them millions of years ago, are particularly rich. The wealth associated with these deposits—the pure rent—is part of a country's geological heritage and presumably belongs to its citizens. Consequently, it is argued, when companies exploit relatively rich deposits, the state should ensure that it captures the pure rent.

A more encompassing variant of this argument is that governments should tax away all of the rents associated with mining and mineral production, since by definition rents can be confiscated without altering the behavior of producing firms and other market actors.

The trouble with both variants of this rationale for mineral taxation lies in the fact, noted earlier, that the rents associated with mining and mineral production—other than possibly monopoly rent—exist only in the short run. While a case might be made for taxing monopoly rents, they are not particularly unique to the mineral sector and are more of an issue for antitrust and competition policy (see Chapter 7). Rather than tax these rents, most economists recommend that governments eliminate the market power that gives rise to such rents.

Regarding the other rents, if governments tax away quasi rents or rents due to supply constraints, mining companies will not earn a competitive rate of return on their investment over the long run. As a result, they will not reinvest in ongoing operations or develop new mines.

Similarly, if governments tax pure rent or reduce rents due to public policy, they diminish the incentives for firms to conduct exploration and to discover new deposits. The hope of finding a bonanza, a deposit so rich it can generate huge amounts of pure rent, drives exploration. In a similar manner, the quest for pure rent motivates the research that creates the new technologies that make previously uneconomic mineral resources now lucrative to exploit. So, unless governments are prepared to conduct the exploration and research needed to create new domestic reserves, those countries that tax away the pure rent and the rent due to public policy face the unpleasant prospect of watching their mineral sector slowly decline over time as their known deposits are depleted.

For this reason, the argument for higher mineral taxes to capture the pure rent or all the Ricardian rent is questionable. Moreover, it is worth highlighting that the presence of large rents in the short run (the sum of the quasi rent, rent due to supply constraint, pure rent, and rent due to public policy) coupled with the absence of these rents in the long run creates a danger of shortsighted policies. Higher taxes on mining and mineral production will almost always raise revenues and appear successful for a time. For this not to be the case, governments would have to confiscate all the short-run rents. The negative effects of higher taxes on mineral production and government revenues may take years to become apparent. Then, unfortunately, they may take many more years to reverse.

2   Non-renewable resources differ from renewable resources in that they have in theory additional value arising from their exhaustible nature, a value measured by their user costs. Since in many countries mineral resources belong to the state, it is often argued that mining companies consuming these resources should compensate the state by paying a tax that covers user costs.

There are, however, two problems with this argument. First, as noted earlier, user costs reflect the value of marginal reserves in the ground, as well as the expected costs of discovering such deposits. So, when governments attempt via taxation or other means to capture user costs, they are actually confiscating assets that mining companies paid for either by undertaking exploration on their own or by purchasing mineral deposits from junior exploration companies or others. Such confiscation inhibits domestic exploration and the discovery of new reserves. This suggests that user costs are best measured by the auction value of the right to explore for mineral deposits rather than the value of mineral deposits after they are discovered, as the latter reflects their expected discovery costs, including the costs of the unsuccessful exploration efforts associated with finding an economic deposit.

Second, there are reasons to believe user costs are negligible for most mineral commodities, including even oil. Chapter 3 noted that the long-run supply curve for individual and joint mineral products at first rises due to the limited number of extraordinary deposits and then becomes much flatter as relatively abundant marginal deposits are exploited. This suggests that the value of marginal reserves in the ground and the cost of discovering such reserves is fairly modest.

Cairns (1981), in an interesting analysis of the Canadian nickel industry, found that user costs accounted for no more than 5 percent, and possibly much less, of the value of nickel production. This finding led him to conclude:

> Given the uncertainties—particularly technological—facing a mining enterprise, it does seem sensible for mining firms not to assume a positive opportunity cost of mining ore now as opposed to in the future.

Pindyck and Rubinfeld (2009), in their microeconomics textbook, have a table (Table 15.7) that reports estimates of user cost as a fraction of the competitive price. For gold, iron ore, nickel, bauxite, copper, and uranium the figures range from 0.05 to 0.30. For oil and natural gas, they are somewhat higher, varying from 0.40 to 0.50. Pindyck and Rubinfeld (p. 577) conclude that "only for crude oil and natural gas is user cost a substantial component of price. For the other resources, it is small and in some cases almost negligible." We suspect even these figures are too high, in part because they include their expected discovery costs. In addition, they may reflect some of the pure rent associated with more attractive deposits. Other empirical studies suggest user costs are more modest, including for oil.[5]

While Hotelling's individual mine must eventually run out of reserves, this is not the case for the world as a whole. Apparently the market does not much worry about the negative effects of production today on the future profits of mining.

Mineral deposits are from time to time sold, and sometimes for substantial sums. In such cases, however, the sale price largely reflects the presence of substantial Ricardian rent. Even when marginal deposits—like Mine G in Figure 5.2, which if developed is expected to just break even—are sold, the price typically reflects an option value, which can be positive, as there is some probability that the market price will rise sufficiently in the future to generate Ricardian rent. In short, it is Ricardian rent and option value—not user costs—that account for all or almost all of the value of mineral resources in the ground.

One reason for this is uncertainty, particularly regarding future technological developments. History shows that saving reserves that can be profitably exploited today for the future is risky. The nitrate deposits of Chile became worthless in the early twentieth century, not because of depletion, but because new technology created attractive substitutes. Similarly, the rise of low-cost ocean transportation for bulk commodities over the past several decades (see Box 4.3) coupled with the discovery of rich iron ore deposits in Australia and Brazil has undermined the economic viability of the iron ore deposits in North America and Europe.

If user costs are insignificant, the value of mineral resources stemming from their exhaustible nature is similarly insignificant. This calls into question the case for raising mineral taxes because mining is exploiting a valuable, non-renewable resource that belongs to the public.

3   A third reason often advanced for raising mineral taxation is the belief that mining companies are simply not paying enough taxes and so are not providing the host country with a fair share of the wealth flowing from its mineral sector. This concern can arise when mineral commodity markets are depressed and governments are receiving little or no tax revenue. It can also arise when mineral commodity markets are booming. While governments reap substantial tax revenues during such times, companies enjoy what appear to be extraordinary profits.

This concern is often found in mining countries, such as Chile, that have relatively low mineral taxes compared to other countries. Does this mean that taxes should be higher in Chile and other states with favorable fiscal regimes? Or that the current share of mineral wealth the government receives is unfair? The answer, unfortunately, is not clear, in part because fairness, like beauty, is a subjective concept over which informed and reasonable people can disagree.

Since all three of these common justifications for mineral taxes have their shortcomings, what can we say about the optimal level of taxation on mining and mineral processing? To address this question, it is helpful to think about this issue in a different way.

As sovereign states, mineral-producing nations should pursue taxation and other policies that achieve, to the maximum extent possible, the goals and objectives they have for their mineral sector. As noted earlier, the objective functions of governments may vary, but a good first approximation is that most countries want to use their mineral wealth to promote the welfare of their citizens. While views may differ over the specifics of this general goal, in the discussion that follows we assume that governments want to pursue taxation policies that maximize the NPV of the stream of tax revenues flowing from the mining sector to the government over time. The government can then use these revenues to foster education, housing, health services, infrastructure, and other goods and services that enhance the welfare of its people.

Of course, the objective function of mineral policy may be more complicated. It may include job creation, regional development, and other considerations in addition to the taxes the mineral sector pays to the government. In this case, the NPV of the social benefits from the mineral sector, which reflects all of the desired goals properly weighted for their relative importance, should replace the NPV of government revenues as the goal or objective of mining taxation policy.

For simplicity, however, we assume in what follows that the government is solely interested in maximizing the NPV of the tax revenues it receives from its mineral sector. In this case, we know that the optimal tax rate, measured as a percentage of profits, is not zero, as this would produce no government revenues. Nor is the tax rate 100 percent, as this would discourage investment and produce zero profits as well.

As Figure 5.3 shows,[6] the NPV of the stream of current and future government revenues increases as the tax rate rises from zero. At some point (T* in Figure 5.3) it reaches a maximum and then declines back to zero as the tax rate approaches 100 percent. Once the optimal is reached, any further increase in the tax rate will cause companies to cut their exploration for new deposits, to stop developing new mines, and eventually, if the tax rate is pushed high enough, to close their existing operations. So raising the tax rate above the optimum at first reduces the fecundity and then kills the goose that lays the golden eggs.

A critical question for many mining countries is whether their current taxes on mining are below or above the optimal rate. If below, mining companies are not paying enough from the country's perspective and the tax level should be higher. Mineral-producing countries have the right, and indeed their governments presumably have the obligation, to pursue policies that ensure their resources contribute the maximum possible to the welfare of their citizens and to society as a whole.

On the other hand, if the current tax rate is above the optimum, attempts to raise it further will be counterproductive. The result will be a reduction in the NPV of tax revenues. In this situation, a lightening of the tax rate is not just good for mining companies; it is also good for the country as a whole.

So we need to know more about the actual shape of the curve shown in Figure 5.3. How does it vary from country to country? And, why? How important, for example, are differences among countries in the perception of their geological potential, their geographic location, and their political stability? Another important question is how the share of rents captured by third parties—organized labor and

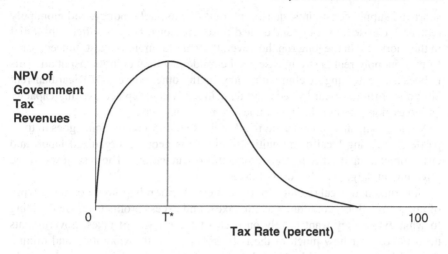

Figure 5.3 The optimal tax rate.

NGOs, for example—affects the curve's shape. In particular, how does the optimal tax rate (T*) and the maximum NPV of tax revenues the government can collect fall as third parties successfully divert more of the rent toward themselves?

We would also like to understand better what causes the curve for any particular country to shift over time. To what extent, for example, are such shifts the result of determinants that the country can control, such as the stability of its tax regime? And to what extent are such shifts outside its control? We know, for example, the curve for Peru will shift to the right, raising the optimal tax rate (T*), when Australia, Canada, and Chile raise the level of their mineral taxes. We still have much to learn, however, about the importance of this and other forces that alter the shape of these curves over time.

Highlights

Mineral economists for the most part favor leaving mineral production in the hands of private companies, believing that they are, on balance, more efficient than state mining enterprises and have better access to global capital markets. In addition, when the state is both the owner and operator, on one hand, and the regulator, on the other, conflicts of interest may compromise its role as regulator.

Most mineral economists also believe the state has an important role to play in the mineral sector. Its active intervention is needed to provide the legal framework for producers, to mitigate environmental pollution and other market failures, and to provide a fiscal regime that best serves the interests of the country.

Mining generates economic rents. Those flowing to mining companies are called Ricardian rents. They reflect a surplus, the loss of which via taxation or other means would not cause mining companies to reduce their production or alter their behavior in other ways. Ricardian rents consist of pure rent, rent due to short-run supply constraints, quasi-rent, rent due to public policy, and monopoly rent, and so arise for various and quite different reasons. They are often substantial in the short run. In the long run, however, they are far more modest. Indeed, aside from monopoly rent, a strong case can be made that they do not exist at all. This is because decreasing or eliminating any of the other sources of Ricardian rent does alter firm behavior by reducing their incentives to replace existing capacity and to explore and develop new mines over the long run.

Mining companies do not capture all of the rent they create. Some goes to third parties, including local communities, indigenous peoples, organized labor, and environmental and other non-governmental organizations. The rest flows to the government, largely in the form of taxes.

Governmental fiscal regimes for the mineral sector rely primary on three types of taxes—royalties, corporate income taxes, and excess profits taxes. In deciding to what extent they want to rely on each of these types of taxes, governments need to consider how much of the risk associated with exploration and mining they wish to assume and how much they wish to leave to mining companies. In addition, they need to take into account the political pressures that arise for changing the rules when the markets are depressed and tax revenues are negligible and as well when markets are booming and private companies are earning extraordinary profits.

Regarding the level of taxation, many contend that producing countries should ensure their mineral taxes capture the pure rent arising from particularly rich domestic deposits. Others stress that metals and other mineral resources are non-renewable and so exhaustible. For this reason they possess an intrinsic value—known as Hotelling rent, scarcity rent, and user costs—that taxes should capture for the country as a whole. Still others simply argue that taxes should be high enough to ensure that the state and producing companies (and perhaps third parties as well) share in a fair manner the benefits flowing from mining and mineral production.

All these justifications for mineral taxes, however, can be questioned for one reason or another. A better rationale is simply that governments should tax and otherwise govern their mineral sectors to serve the interests and welfare of their citizens. This requires that governments maximize the NPV of current and future mineral taxes or, more broadly, that they maximize the NPV of the social benefits flowing from mining.

The NPV of the social benefits flowing from mining varies with the tax rate. The benefits follow a curve that rises, reaches a peak, and then falls as the tax rate on profits increases from 0 to 100 percent. The optimal tax rate produces the greatest level of benefits.

The curve and the optimal rate vary from country to country and over time for individual countries as well. Unfortunately, in practice it is not easy to estimate either the curve or the optimal rate. More troubling is the fact that when mineral taxes are raised, government revenues almost always increase for a time. It takes years for the negative effects of higher taxes to become apparent.

Despite such difficulties, however, governments should strive to set the level of their mineral taxes so that their objective function reflecting the interests of the country is best served. When the tax rate is below the optimal level, governments have an obligation to raise the rate. When it is above the optimal level, governments should lower the rate.

Notes

1   This and the following sections of this chapter draw extensively from Tilton (2004) and the sources noted there.
2   One caveat is perhaps worth noting. If some of the million dollars the individual receives for playing soccer is compensation for the fact that his career is likely to be limited to 15–20 years (while mineral economists can continue working much longer), then the rent he earns is less than $800,000.
3   Sometimes Ricardian rent is defined as solely pure rent. However, we define Ricardian rent more broadly to include all of the economic rent flowing to mining companies.
4   The point being made here, it is worth highlighting, does not question the desirability of government regulations designed to force firms to pay for pollution and the other social costs associated with their production.
5   See, for example, Adelman (1990), Adelman and Watkins (2005), Cairns (1998), Halvorsen and Smith (1984; 1991), and Kay and Mirrlees (1975).
6   The curve in Figure 5.3 is a modified Laffer curve. The latter shows the relationship between the tax rate and tax revenues. It highlights the fact that as the tax rate rises revenues will first rise and then fall (as higher rates eventually reduce the incentives for people to work). The curve was proposed by the US economist Arthur Laffer.

References

Adelman, M.A., 1990. Mineral depletion with special reference to petroleum. *Review of Economics and Statistics*, 72 (1), 1–10.
Adelman, M.A. and Watkins, G.C., 2005. U.S. oil and natural gas reserve prices, 1983–2003. *Energy Economics*, 27, 553–571.
Cairns, R.D., 1981. An application of depletion theory to a base metal: Canadian nickel. *Canadian Journal of Economics*, 14 (4), 635–648.
Cairns, R.D., 1998. Are mineral deposits valuable? A reconciliation of theory and practice. *Resources Policy*, 24 (1), 19–24.
Halvorsen, R. and Smith, T.R., 1984. On measuring natural resource scarcity. *Journal of Political Economy*, 92 (5), 954–964.
Halvorsen, R. and Smith, T.R., 1991. A test of the theory of exhaustible resources. *Quarterly Journal of Economics*, 106 (1), 123–140.
Hotelling, H., 1931. The economics of exhaustible resources. *Journal of Political Economy*, 39 (2), 137–175.
Kay, J.A. and Mirrlees, J.A., 1975. On comparing monopoly and competition in exhaustible resource exploitation, in Pearce, D.W., ed., *The Economics of Natural Resource Depletion*, Macmillan, London, 140–176.
Mikesell, R.F., 1975. *Foreign Investment in Copper Mining: Case Studies of Mines in Peru and Papua New Guinea*, Resources for the Future, Washington, DC.
Otto, J., Andrews, C., Cawood, F., Doggett, M., Guj, P., Stermole, F., and Tilton, J., 2006. *Mining Royalties: A Global Study of their Impact on Investors, Government, and Civil Society*, World Bank, Washington, DC.

Pindyck, R.S. and Rubinfeld, D.L., 2009. *Microeconomics*, seventh edition, Pearson
  Prentice Hall, London.
Tilton, J.E., 2004. Determining the optimal tax on mining. *Natural Resources Forum*, 28,
  144–149.

6 Mineral Commodity Trade and Comparative Advantage in Mining

Two hundred years ago most of the world's production and consumption of mineral commodities took place in Europe and the United States. Over time, as their best deposits were depleted and the cost of shipping bulk commodities declined, these areas increasingly relied upon imports, often from developing countries. More recently, as pointed out in Chapter 2, Japan and then China have become major consumers, relying heavily on imports for many of their mineral resource needs.

The growing importance of international mineral trade raises a number of interesting questions. For instance, how stable over time is comparative advantage in mining? (See Box 6.1 for definitions of comparative advantage and competitiveness.) Are the major producing countries today the same as 25 or 50 years ago? Where there have been changes, why did they occur? Is mineral depletion largely the answer or are there other important causes? More generally, what are the major determinants of comparative advantage in mining? How have they changed over time? What have been the consequences? And what are the important policy implications for mineral-producing firms and countries?

This chapter explores these questions. It begins by reviewing recent trends in the location of mining for the more important mineral commodities. It then examines international trade theory and in particular the doctrine of comparative advantage. This leads to an analysis of mineral endowment and the other important determinants of competitiveness in mining. Finally, the chapter explores why comparative advantage in mining shifts over time from one set of countries to another and the policy implications for producing firms and countries.

We assume a country possesses competitiveness and a comparative advantage in mining if it is a major ore producer. Usually this means that the country enjoys relatively low production costs, but this need not be the case. For example, a country may produce a lot because its government subsidizes domestic firms or protects them from low-cost imports. Alternatively, a country may be a major producer because it developed its mine capacity—which can last for decades—during an earlier period when its costs were lower.

Thus, the term comparative advantage is used here in a broader sense than in many other studies. In particular, international trade theory, which we examine later in this chapter, generally assumes that comparative advantage depends solely

Box 6.1 Competitiveness and Comparative Advantage

Competitive and competitiveness have two different meanings. When we refer to a competitive industry, for example, we may mean an industry where no producer or other market actor possesses the ability and the incentive to alter the market price. This is the definition of competitiveness that Chapter 4 uses in describing mineral commodity markets.

Alternatively, we at times refer to an industry—the Brazilian iron ore industry, for example—as being competitive when we mean that it accounts for a large share of the global production thanks to relatively low production costs or other advantages. In this sense of the word, the industry is competitive—or possesses competitiveness—because its producers can compete successfully against producers elsewhere, and as a result enjoy a significant share of the market.

The term *comparative advantage* as used in this volume has the same meaning as this second definition of competitiveness. It implies that a country and its producers are strong competitors and account for a significant share of total production reaching the market. As the text notes, however, some studies define comparative advantage more narrowly to reflect a country's share of global exports rather than global production.

on comparative costs. Indeed, the two terms are often used as synonyms in this literature. In addition, trade theory normally assumes that comparative advantage leads to a large share of global exports, rather than a large share of global production. Of course, the two are often closely correlated, but this does not have to be the case. The United States and China, though both large producers of copper, are net importers, not exporters, of this metal.

In this chapter, however, we are ultimately concerned with the ability of countries to produce mineral commodities for both export and domestic consumption. For this reason, a broader definition of comparative advantage—one that reflects shares of global production—is used.

Patterns and Trends in Mine Production

Table 6.1 identifies the top three producing countries for six important mineral commodities for the years 1960, 1985, and 2010. Over this period spanning half a century, the global output for all these commodities expanded rapidly. The table highlights the frequent changes among the top three producing countries. Comparative advantage in mining, it appears, is a fickle friend.

Bauxite provides a particularly dramatic example. In 1960, Jamaica, the Soviet Union, and Suriname were the largest producers, accounting for almost half of world output. By 1985, Jamaica had fallen from first to third place, and Australia and Guinea had replaced the Soviet Union and Suriname in the top tier. Twenty-five years later, in 2010, Australia remained the top producer but was now followed by China and Brazil.

Table 6.1 The three largest producing countries for bauxite, copper, iron ore, nickel, phosphate rock, and potash in 1960, 1985, and 2010 with their shares of world production[a]

	1960	1985	2010
Bauxite	Jamaica (22)	Australia (36)	Australia (33)
	USSR (12)	Guinea (17)	China (19)
	Suriname (11)	Jamaica (7)	Brazil (15)
Copper	USA (23)	Chile (28)	Chile (34)
	Zambia (14)	USA (22)	Peru (8)
	Chile (12)	USSR (21)	China (7)
Iron ore	USSR (25)	USSR (31)	China (37)
	USA (19)	Brazil (13)	Australia (17)
	France (9)	Australia (10)	Brazil (15)
Nickel	Canada (57)	USSR (23)	Russia (17)
	USSR (17)	Canada (22)	Indonesia (15)
	New Caledonia (16)	Australia (11)	Philippines (10)
Phosphate rock	USA (44)	USA (33)	China (38)
	Morocco (18)	USSR (21)	USA (14)
	USSR (16)	Morocco[b] (14)	Morocco[b] (14)
Potash	Germany (46)	USSR (35)	Canada (29)
	USA (30)	Canada (23)	Russia (19)
	France (19)	Germany (21)	Belarus (16)

Sources: US Geological Survey (annual); World Bureau of Metal Statistics (annual).

Notes:
a   Figures in parentheses indicate a country's share of world production.
b   Includes Western Sahara.

Frequent changes occurred as well among the leading producers of the other commodities considered. Of course, given the growth in global output, a falling market share does not necessarily imply an actual decline in a country's output.

Another striking development is the rise of China as a mining power in recent years. In 1985, China was not among the top three producers for any of the mineral commodities considered. By 2010, it was the world's largest producer of iron ore and phosphate rock, the second largest producer of bauxite, and the third largest producer of copper. We tend to think of China as an emerging consumer of raw materials, but its growth as a producer for many mineral commodities is also impressive.

Aside from the rise of China, Table 6.1 does not provide much evidence that mining is shifting dramatically toward developing countries. Australia, Canada, and the United States remain important producers for a number of the major mineral commodities. In addition, some of the emerging country producers—Chile, China, and Brazil, for example—are enjoying rapid economic development. Indeed, Chile recently joined the OECD (the Organization of Economic Co-operation and Development), whose members are widely considered developed countries.

Another trend that Table 6.1 reflects for bauxite and iron ore is a rising share of total production captured by the top three countries. This mostly reflects three developments that have reduced barriers to trade and allowed mining to shift to countries with the best deposits and the lowest production costs—the falling costs of shipping mineral commodities; the reduction in tariffs and other trade restrictions under the auspices of GATT (the General Agreement on Tariffs and Trades) and its successor the WTO (World Trade Organization); and the integration of Eastern Europe, China, and the former member states of the Soviet Union into the global economy over the past several decades.

Finally, it is important to underline that competitiveness in mining does not necessarily translate into competitiveness in downstream processing. While the three top producers of bauxite in 2010 were Australia, China, and Brazil, in that order, the three top aluminum producers were China, Russia, and Canada. Neither Canada nor Russia is an important producer of bauxite.

What explains this weak association between comparative advantage in mining and comparative advantage in downstream stages of production? In the case of bauxite and aluminum, energy—especially electric power—accounts for a much higher portion of the total costs of producing aluminum than bauxite. As a result, countries with low-cost electric power, even though they have to import bauxite, often enjoy lower aluminum production costs than most bauxite producing countries. In other situations, importing countries, including some major consumers such as China and Japan, encourage the domestic processing of metals and other mineral commodities by imposing lower tariffs on imports of ores and other less processed mineral commodities.

Of the various issues that Table 6.1 raises, perhaps the most perplexing is why the market shares of the major producing countries have so often changed, at times abruptly and substantially, over the past half-century. The quest for an answer begins in the next section with a brief examination of international trade theory.

International Trade Theory

The reasons why some countries produce and export certain goods while other countries import these goods and export others have interested economists for centuries. Modern explanations found in the pure theory of international trade are based on the *doctrine of comparative advantage*, also known as the *doctrine of comparative costs*, proposed by the classical economist David Ricardo some two centuries ago. This doctrine maintains that countries will produce and export those commodities whose domestic costs of production are low relative to their domestic costs of production for other products when compared with production costs in other countries.

For example, in the absence of trade, assume that in Australia a basket of primary commodities costs 100 Australian dollars to produce and a basket of manufactured products 200 Australian dollars. In China, the same basket of commodities costs 800 yuan and the same basket of manufactured products 1200 yuan. Without trade, two baskets of commodities would exchange for one basket of manufactured products in Australia. In China, the basket of manufactured products would be cheaper, costing only one and a half baskets of commodities.

According to the doctrine of comparative advantage, Australia should as a result export commodities to China, and China export manufactured products to Australia. Trade between the two would benefit both countries even though one might be more efficient and so possess an absolute advantage in the production of both sets of goods (in the sense that their production requires less capital, labor, and other inputs per unit of output). All that trade requires is that the exchange rate between the Australian dollar and the yuan is set so that it is cheaper for Australia to buy manufactured products from China than to produce them domestically (which occurs when the Australian dollar is worth more than six yuan), and cheaper for China to buy commodities from Australia than to produce them domestically (which occurs when the Australian dollar is less than eight yuan).

The doctrine of comparative advantage is just the first step in explaining trade. For it to be useful, we need to know what accounts for differences in comparative costs. Why in the example above does Australia enjoy a comparative cost advantage in commodities and China in manufactured goods?

David Ricardo and the classical economists were the first to answer this question, maintaining that differences in labor productivity were responsible. A farmer in Australia might produce three times the amount of wheat or mutton in a day as a farmer in China because of more favorable terrain, larger-scale operations, more highly mechanized techniques, and other advantages. In comparison, Australian workers making manufactured goods might produce only 50 percent more than their Chinese counterparts.

While labor productivity helps explain some of the differences in production costs, there are other important factors of production besides labor. In the twentieth century, Jacob Viner, Gottfried Haberler, and other neo-classical writers tried to rectify this shortcoming of the classical explanation by suggesting that differences in total productivity, rather than just labor productivity, lie behind international differences in comparative costs.[1] Differences in total productivity arise, they suggest, because production functions for goods vary from one country to another. The Chinese may apply the same amount of labor, machinery, and fertilizer to their agricultural sector and still not reap the same output per acre as in Australia.

The *factor endowment theory*, advanced by the Swedish economists Eli Hechscher and Bertil Ohlin in the early years of the twentieth century, offers an alternative explanation for differences in comparative costs. In contrast to the neo-classical theory, it maintains that production functions—which reflect the technical relationships that govern how much output can be produced with various quantities of inputs—are the same throughout the world. International differences in production costs arise instead because countries possess different factor endowments, which in turn create differences in input prices. Australia, for example, is well endowed with good land for farming and grazing; China with people and capital. As a result, the price of good agricultural land is lower in Australia, while the price of labor and capital is lower in China. So Australia enjoys a comparative advantage in the production of goods, such as food, that are land intensive, and China a comparative advantage in the production of goods that are labor and capital intensive, such as manufactured products.

During the latter half of the twentieth century, other explanations for comparative advantage emerged. These theories stress the importance of inter-country differences in technology, human capital, economies of scale, and consumer preferences. They arose primarily to explain the large volume of trade that takes place between developed countries in manufactured products. While differences in factor endowments and productivity do exist among the developed countries, they typically are small compared with those found between the developed and developing countries. Moreover, developed countries with apparently similar domestic conditions, such as those of Western Europe, often import substantial quantities of similar finished goods from each other. This has raised concern over the usefulness of the neo-classical and factor endowment theories in explaining trade between such countries in automobiles, computer software, and other final products.

This dissatisfaction, however, does not extend to trade in mineral commodities and other primary products. Here, it is still widely accepted that the factor endowment theory is the most useful explanation for international differences in production costs. In the case of metals and other mineral products, the applicability of the factor endowment theory appears almost self-evident. In the words of Haberler (1977, p. 4):

> The most obvious factors that explain a good deal of international trade are "natural resources"—land of different quality (including climatic conditions), mineral deposits, etc. No sophisticated theory is required to explain why Kuwait exports oil, Bolivia tin, Brazil coffee and Portugal wine. Because of the deceptive obviousness of many of these cases, economists have spent comparatively little time on "natural resource trade."

So Australia and Brazil produce and export bauxite and iron ore because they are well endowed with these resources. Intuitively, at least, the hypothesis that resource endowment is the overriding determinant of comparative advantage in mineral production is very appealing. The next section examines this idea more closely.

Mineral Endowment and Other Determinants

Chile is the world's leading producer of copper. Presumably this country is well endowed with copper, but just how much copper does it have? And how does its endowment compare with those of Peru, China, and other copper-producing countries? More generally, is there a close relationship between a country's endowment of any mineral commodity and its mine production, as the factor endowment theory anticipates?

Measuring Mineral Endowment

Before we can address such questions, we need a measure for mineral endowment. Mineral reserves are probably best for this purpose. As noted earlier, reserves measure the amount of a metal or other mineral commodity contained in deposits that are both known (discovered) and profitable to mine given current prices and production costs. Other possible measures of endowment, such as mineral resources (which measure the amount of a commodity not only in deposits that are currently known and profitable but also those anticipated to be so in the future) may influence the future. However, mine production can take place only after mineral deposits are developed, which requires that they be both discovered and considered profitable.

It is important to highlight, however, that reserves are not fixed but change over time as new discoveries occur, new technologies develop, and other conditions vary. As a consequence, to the extent comparative advantage in mining rests on mineral endowment as measured by reserves, it can shift from one set of countries to another over time. So finding a strong relationship between reserves and mine output just raises another question: What forces determine country reserves and cause them to rise or fall over time? We return to this important question below, but first we need to explore just how much of the differences in comparative advantage in mining (as measured by mine output) can be explained by differences in country endowments (as measured by their reserves).

Reserves and Mine Output

Normally a decade or more elapses between the discovery of a new deposit and the start of production. This time is needed to conduct feasibility studies, prepare the environmental impact statements, plan the development, obtain the necessary permits, arrange the financing, deal with legal challenges raised by NGOs and others, strip the overburden for an open-pit mine or sink the shafts for an underground mine, erect concentrators, construct transportation facilities, and carry out numerous other steps, often involved in bringing a new mine on-stream. If a mine is located in a remote region of a developing country, it may be necessary to build a town site and port facilities, and perhaps to work with the host government as a partner in the project.

The lengthy gestation period between discovery and first production means that in examining the relationship between mine output and reserves we should lag reserves by a number of years. This is done in Figure 6.1, which regresses copper production for the major producing countries in 1960, 1985, and 2010 on their reserves a decade earlier.

In examining this figure, we should keep in mind that the data on reserves, though they have improved considerably over the past half-century, are still far from perfect. Just how reserves are defined and measured can vary from one country to another and even from one deposit to another. This is probably less of a problem for copper than for other mineral commodities, such as iron ore, found in a large number of deposits and countries. In the case of iron ore, another concern

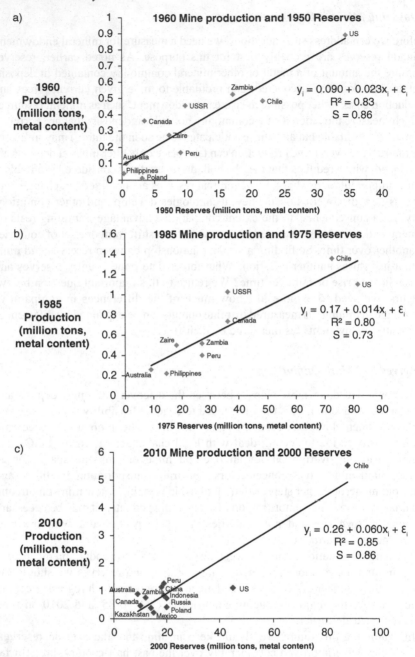

Figure 6.1 Copper mine production regressed on reserves lagged ten years: (a) 1960 mine
         production and 1950 reserves; (b) 1985 mine production and 1975 reserves;
         (c) 2010 mine production and 2000 reserves (sources: see Tables 6.1 and 6.2,
         and the sources cited there).

is the difference in the metal content of reported reserves. A ton of ore in Australia or Brazil, for example, typically contains twice the amount of iron as a ton of ore in China. Fortunately, such differences in ore grades are much smaller for most other mineral commodities. Moreover, reserves are often measured in terms of content of the commodity in the ore—for example, the tons of actual copper contained in the ore—rather than in the tons of ore itself.

Figure 6.1 identifies the major copper-producing countries and their reserves for each of three time periods. It shows the linear relationship between the production (y_i) of these countries and their reserves (x_i), determined by simple regression analysis along with the estimated equation. All of the coefficients for the reserves variable (x_i) are positive and statistically significant (meaning that the coefficient on the reserve variable is greater than zero—the value we would expect if reserves have no influence on production—with 95 percent probability).

For each of the three equations, Figure 6.1 also reports the coefficient of determination (R^2), which shows how well each equation fits the data points used in its estimation. This statistic indicates how much of the variation in output among the top producing countries might be attributable to or explained by differences in their reserves. As with all such regressions, however, the estimated relationships between two variables simply show an association or correlation, which may or may not reflect cause and effect. Finally, Figure 6.1 reports for each equation the share (S) of world output accounted for by the identified countries.

We created similar figures for other important metals, namely bauxite, iron ore, lead, nickel, and zinc. These figures are not reproduced here (in the interest of space), and there is no need to explore in detail all the findings. Instead, we examine in depth the relationship between copper production in 2010 and reserves in 2000, and then highlight the important general conclusions that emerge when the results for the other metals and time periods are taken into account.

Chile was by far the largest copper-mining country in 2010, followed by Peru, China, the United States, Australia, Indonesia, Zambia, and Russia. The four other countries that Figure 6.1 considers are Canada, Poland, Kazakhstan, and Mexico. Together, these countries account for 86 percent of the 2010 world copper output. The estimated relationship between their production and reserves, shown by the solid line in the bottom panel, is $y_i = 0.26 + 0.060\,x_i + \varepsilon_i$, where y_i is the production of the ith country measured in millions of tons and x_i its reserves, also measured in millions of tons. ε_i is the error term, which among other things captures the influence of variables affecting production other than reserves. The coefficient on the x_i variable (0.060) implies that a one million ton increase in 2000 reserves is associated with a 60,000 ton increase in 2010 production.[2] As noted earlier, this coefficient is statistically significant. So there is no more than 1 chance in 20 that the estimated positive relationship between reserves and production is due to mere chance. Finally, the coefficient of determination (R^2) is 0.85, which suggests that differences in reserves can account for or explain up to 85 percent of the variation in output among the major copper-producing countries.

The findings for other time periods for copper as well as for the other metals analyzed support two particularly important generalizations or conclusions. First, in all but two instances (nickel and zinc production for 1960), a positive relationship exists between output and reserves lagged a decade. In many (though not all) cases this relationship is statistically significant. Over time the relationship between output and reserves has remained more or less the same or grown stronger for all the metals considered, with the exception of bauxite and iron ore. Overall, these findings provide some support for the factor endowment theory of comparative advantage in mining. As a country's reserves rise, so too do its expected production and share of world output.

Second, the estimated relationships between country production and reserves are far from perfect. In 2010, for example, Chile, Peru, and Australia mined more copper than expected solely on the basis of their reserves, while the United States, Poland, and Mexico produced less. In a few instances, the coefficients of determination (R^2) indicate that reserves can explain more than 80 percent of the variation in country output. In most cases, however, the percentage is lower, presumably reflecting the existence of other important determinants of comparative advantage besides reserves and mineral endowment.

Other Factors Affecting Comparative Advantage

Mining, of course, entails more than just good deposits. The drills, trucks, crushers, mines, mills, infrastructure, and other necessary facilities require capital, usually in large amounts. In addition, labor is needed to run and maintain equipment, and to provide technical and managerial skills. Energy and material supplies are also essential.

However, these inputs—with the important exception of cheap electric power—are mobile. Major mining firms have for many years been willing to take their capital along with their managerial and technical expertise to remote corners of the world, as long as the investment environment was otherwise favorable. As a result, international differences in the endowment of capital and expertise are of secondary importance in the location of mining facilities. Of greater significance are the legacy of past investment and the political environment.

Legacy of Past Investment

Much of the capital invested in a mine cannot be recovered by closing down the facility. So it pays to continue operating an existing mine, rather than develop a new one, as long as the average out-of-pocket or variable costs of production, including the costs of necessary repairs, are lower than the average total costs of a new mine. Since capital costs are normally a large part of total costs, once a mine is constructed, it will typically remain in operation until the associated ore body is depleted.

If the life expectancy of most mines were short, this would not greatly delay geographic shifts in production and comparative advantage. But, as is well known, this is not the case. Most major mines remain in operation for at least 20 to 30 years, and some carry on for many decades more. Bingham Canyon in the western

United States and Chuquicamata in northern Chile were both developed in the early 1900s and are still two of the biggest operating copper mines in the world.

The long, productive lives of mines coupled with their high capital costs can prolong the dominance of established mining districts long after the conditions that promoted their rise cease to exist. Two other considerations also favor established mining districts. First, new mines can usually be developed more cheaply in areas where adequate transport systems and other infrastructure are already in place and where skilled labor is readily available. Second, and somewhat related, the cost of expanding existing mines is often less than building new ones. At existing facilities, administrative and overhead costs typically do not rise in proportion to the increase in capacity. This may apply to maintenance costs as well. In addition, mining companies can in some instances increase capacity with minor investments that alleviate bottlenecks in the production process.

The benefits of exploiting operating mines until they are exhausted, of developing new mines in established mining districts, and of adding new capacity by expanding existing facilities can help explain why in 2010 Australia mined more bauxite and copper and why Brazil mined more iron ore than expected on the basis of their reserves.

Political Environment

The political environment in mineral-producing countries has both a direct and indirect effect on their output and comparative advantage in mining. The indirect effect, considered in the next section, arises because government policies and actions influence the geographic distribution of exploration activity. This in turn affects the location of new reserves and ultimately comparative advantage in mining. The direct effect, on the other hand, causes mining countries to produce more or less than expected on the basis of their reserves.

The direct effect occurs in part because political considerations—including heavy taxation, exchange controls, environmental regulations, lengthy permitting procedures, requirements to use domestic services and goods, demands to process ores domestically, and corruption—can make even the highest-quality resources too expensive to develop. In addition, public policies that promote widespread domestic discontent while suppressing peaceful change, or that incite tensions and hostility with neighboring countries, may encourage civil disruptions and war, leaving good domestic deposits too risky to exploit.

Examples of public policies adversely affecting mine production and comparative advantage are quite numerous and easy to identify. Clearly, a more favorable political climate in Russia and a number of African countries (including the Democratic Republic of the Congo and Zimbabwe) since 1990 would have encouraged more foreign investment in their mining sectors. During the 1960s and 1970s, as Chapter 3 noted, the tendency to nationalize mining properties led to a dearth of new investment and a decline in production in many developing countries.

The effects of the political environment—both direct and indirect—are not always negative. Governments can promote mining by constructing needed infrastructure, granting generous tax holidays, providing inexpensive electric power, and offering other benefits. For example, the Chilean government encouraged the rapid expansion of its copper industry during the 1990s with deferred tax provisions and other favorable policies. China today mines more bauxite, copper, iron ore, lead, and zinc than expected solely on the basis of its reserves. This, in large part, reflects the influence of public policies favoring domestic mining, including the ready availability of low-cost capital as well as tariffs on many mineral commodity imports.

Forces for Change[3]

The preceding section provides insights into the nature and determinants of comparative advantage in mining. But, it still leaves unanswered the question of why comparative advantage in mining shifts over time, causing dramatic changes in the top mining countries for many mineral commodities. The association found between country production and reserves lagged ten years suggests that changes in reserves—and the forces behind such changes—may account for many of the shifts observed in comparative advantage.

Changes in Reserves

Table 6.2 identifies the three countries with the largest shares of global reserves for bauxite, copper, iron ore, and nickel for the years 1950, 1975, and 2000. We selected these three years because they are a decade prior to the three years of mine output that Table 6.1 considers, and so are the reserves figures used in estimating the regressions of mine output on reserves for copper (reported in Figure 6.1) and for the other metals.

Not surprisingly, Table 6.2 shows that the top reserve-holding countries have changed dramatically over time, just like the top mining countries examined a decade later in Table 6.1. In the case of bauxite, for example, Guinea, Brazil, and Australia together held 60 percent of the world's reserves in the year 2000 and none or almost none in 1950. In the case of iron ore reserves, Australia was second only to Ukraine in 2010. However, its reserves in 1950 were so meager that the country prohibited iron ore exports on the grounds that all domestic supply should be saved for the exclusive use of the small Australian steel industry.

So, if we want to understand why comparative advantage in mining shifts over time, we need to understand the forces driving changes in country reserves. Here, three possible explanations come readily to mind. First, as mining occurs and the best (lowest cost) deposits are depleted, mineral commodity prices rise. This permits the profitable exploitation of the next best set of deposits, which may be located in different countries than the earlier and now exhausted reserves. Second, exploration and new discoveries can add to existing reserves. Third, innovation and new technology can create reserves by permitting the profitable exploitation of formerly known but uneconomic deposits.

Table 6.2 The three countries with the largest reserves for bauxite, copper, iron ore, and nickel in 1950, 1975, and 2000 with their shares of total world reserves[a]

	1950	1975	2000
Bauxite	Jamaica (18)	Guinea (30)	Guinea (29)
	Guyana (5)	Australia (26)	Brazil (16)
	France (4)	Brazil (10)	Australia (15)
Copper	USA (29)	USA (22)	Chile (28)
	Chile (19)	Chile (19)	USA (14)
	Zambia (15)	Canada (10)	Poland (6)
Iron Ore	India (21)	USSR (31)	Ukraine (20)
	Brazil (16)	Brazil (16)	Australia (17)
	France (9)	Canada (12)	Russia (17)
Nickel	Cuba (79)	Cuba (22)	Cuba (15)
	Canada (15)	New Caledonia (21)	Australia (12)
	New Caledonia (4)	Canada (13)	Canada (10)

Sources: US Geological Survey (annual); US Bureau of Mines (1952a, 1952b, 1953, 1976); United Nations (1955).

Note:
a   Figures in parentheses indicate a country's share of world reserves.

Despite the rise in prices for many metals and non-metallics over the past decade, the first possible explanation for changes in country reserves is probably the least important of the three. This is because over the past 100–150 years, despite quite strong cyclical fluctuations over the short run, long-run trends in real prices for most mineral commodities have not been rising.

We do know that the second explanation—the discovery of previously unknown deposits—is important and can cause dramatic shifts in country reserves. But, is it as important as the third explanation and the new technologies that permit the profitable mining of previously known but uneconomic resources?

On this question, one finds two different perspectives with very different implications for mineral-producing firms and countries. The traditional view of mineral economists and other analysts stresses the importance of the first two explanations and, for the most part, ignores the third. The alternative view, on the other hand, focuses on the third explanation. It contends that innovation and new technology—not price changes and new discoveries—are largely responsible for the creation of new reserves and competitiveness in mining, particularly over the longer term.

While the jury is still out as to which of these two perspectives is most valid, growing evidence supports the alternative view and the importance of innovation and new technologies. This evidence includes studies of the US and Chilean copper industries, which one of us (John Tilton) conducted with various collaborators in the late 1990s and early 2000s. We delve next into the highlights of this work.

US Copper Industry, 1975–2004

Throughout much of the twentieth century, the United States was the world's leading copper producer. By the late 1970s and early 1980s, however, its industry was in trouble. Between 1970 and 1985, output declined by nearly one-third and its share of Western world production fell from 30 percent to 17 percent. Employment dropped by 70 percent. Out-of-pocket or cash costs declined, but not enough to keep pace with the drop in market price. As a result, in the early 1980s few mines were earning a profit, and many were not even covering their variable or cash costs.

US copper producers petitioned the government for protection from imports in 1978 and again in 1984, claiming their very survival was at stake. On both occasions, their request was denied. The media, as well, lamented the industry's fortunes. *Business Week* in the mid-1980s ran a cover story declaring the death of mining in the United States.

Amoco Minerals, Arco/Anaconda, Cities Service, Louisiana Land and Exploration and other companies left the industry. They sold their mines to other companies, spun them off as independent companies, or simply shut them down. Of the 24 significant copper mines operating in the United States in 1975, six had closed by 1990 and another five had sharply cut back production.

Yet the industry did survive, staging one of the most spectacular turnarounds in modern industrial history. By 1995, output was 72 percent above its 1985 level and even 21 percent above its 1970 level. Its share of the Western world market recovered to 23 percent. Imports were down and profits up as costs continued to fall while prices recovered.

Innovation, Mineral Endowment and Comparative Advantage

Many factors contributed to the recovery of the US copper mining industry. They included a decline in real wages, an increase in by-product revenues, a rise in copper prices and the depreciation of the dollar. Among these, however, a dramatic improvement in labor productivity was more equal than others. As Figure 6.2 shows, labor productivity more than doubled between 1980 and 1986. So, where two workers were needed in 1980, one would do six years later. Labor productivity continued to rise after 1986, though at a more modest pace. Still, by 2001 it was three times its 1980 level.

Part of this surge in labor productivity can be attributed to an increase in the amount of capital, energy, and other factors available per worker. During the 1980s, for example, Bingham Canyon undertook a 400 million dollar modernization program that helped the mine quadruple its labor productivity. Even more important, however, was the introduction of new technologies and other innovations.

One important development was the increasing use of the solvent extraction electrowinning (SX-EW) process (see Box 6.2). This technology greatly reduced the operating and capital costs of producing copper. A better understanding of rock mechanics allowed new mine plans that reduced stripping ratios and so diminished the amount of waste generated per ton of ore. Innovative agreements

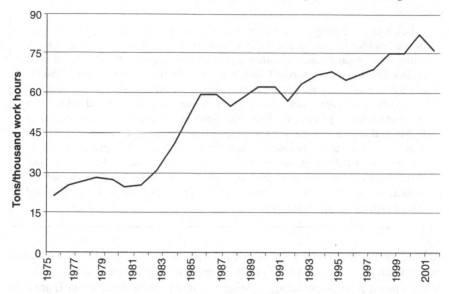

Figure 6.2 Labor productivity in the US copper industry, 1975–2001 (tons of copper contained in mine output per 1,000 man-hours by copper company employees) (sources: US Geological Survey, US Mine Safety and Health Administration, as cited in Tilton (2003)).

with labor increased the flexibility in work rules and manning assignments. Other innovations included better ore handling systems, larger trucks and shovels, bigger drills, in-pit mobile crushers with conveyor belts, more cost-effective explosives, and the computerization of truck schedules and real-time process controls in mills.

Box 6.2 The Solvent Extraction and Electrowinning Technology

The traditional technology for producing copper entails mining sulfide copper ore in underground or open-pit mines. The ore is then moved by truck, rail, or conveyor belt to a mill. Here, it is crushed and the copper-bearing minerals are separated from the waste material or gangue by flotation. The resulting concentrate (25–35 percent copper) is transported to a smelter for partial purification (97–99 percent copper) and then on to a refinery for electrolytic purification (99.99 percent copper).

The SX-EW process for producing refined copper relies on a more recent and quite different technology that extracts copper primarily from oxide rather than sulfide minerals. The first step involves leaching existing mine dumps, prepared ore heaps, or in situ ore with a weak acidic solution. The solution is recovered and, in the next stage, the solvent extraction stage, mixed with an organic solvent (referred to as an extractant), which removes the copper from the solution. The copper-rich extractant is then mixed with an aqueous acid solution, which strips it of its copper. The resulting electrolyte is highly concentrated and relatively pure. It is processed into high-quality copper in the third and final stage by electrowinning.

Leaching mine dumps from earlier operations does not require mining. Nor does in situ leaching, where copper is extracted from fractured ore remaining in place within the original deposit. Leaching of prepared ore heaps, on the other hand, does require that the ore be removed from the deposit and in some cases crushed. So the SX-EW process often, though not always, avoids the substantial costs of mining.

SX-EW is a hydrometallurgical process, while the traditional technology is a pyrometallurgical process as it entails smelting. There are, however, other hydrometallurgical processes for treating copper ores. For centuries, copper has been leached and then recovered by precipitation with iron. This process produces cement copper (85–90 percent copper), which then must be smelted and refined. Direct electrowinning involves leaching and electrowinning, but skips the solvent extraction stage. It produces a lower-quality copper contaminated with iron and zinc, which also requires further processing.

Relying on the traditional view of comparative advantage in mining, we would expect to find behind the revival of the US copper mining industry an improvement in the mineral endowment being exploited—either from raising the cutoff grade at existing mines or from shifting production from high-cost to low-cost mines. But there is not much evidence of either.

Trends in copper head grades do show a rise in the early 1980s, from 0.59 percent in 1980 to 0.68 percent in 1984. Some mines with poorer deposits closed and other mines turned to higher-grade ores to reduce their costs during this particularly difficult period. However, the rise in head grades was short lived. And over the entire 1971–1993 period grades drop considerably, from 0.78 percent to 0.60 percent (Tilton and Landsberg, 1999, fig. 4.5).

We also know that shifts in mine location did not play a dominant role. The new mines brought on-stream during the 1975–1990 period, including Flambeau and Cyprus Tohono, contributed little to the country's output, less than 5 percent. So the revival of the US industry came about because existing mines recovered their competitiveness. In particular, Bagdad, Chino, Morenci, Ray, and Tyrone more than doubled their output, while Bingham Canyon increased its production by 50 percent.

These substantial increases raise the possibility that the revival of the US industry was largely the result of productivity improvements and cost reductions flowing from a shift in output away from poor high-cost deposits to the good deposits at these mines. However, measuring how much of the rise in labor productivity for the industry was the result of shifts in output from low- to high-productivity mines, and how much was the result of individual mines increasing their productivity, we find that the shift in mine location accounted for only one-quarter of the rise in industry productivity (Aydin and Tilton, 2000). This means that three-quarters of the total increase came about as a result of improvements in labor productivity at individual mines, where mineral endowments presumably changed little. These findings suggest that changes in mineral endowment were of secondary importance compared with innovative activity in the recovery of the US industry.

Technology Diffusion and Comparative Advantage

The US copper mining industry challenges the traditional view of comparative advantage in another way. As noted earlier, adherents of the traditional view claim that innovation and new technology have little or no influence on competitiveness. This is because new technology in the global economy diffuses rapidly around the world. For example, they contend, there is little or no difference in the time at which a new and more efficient shovel or explosive is available to mines in the United States, Chile, Zambia, or elsewhere. So a cost advantage based on new technology either will not arise at all or will be short-lived.

This conclusion, however, rests on two implicit assumptions. The first is that a new process or technique is the result of but one innovation. The second is that its effects are neutral in the sense the impact on costs is the same for all producers. As the SX-EW process illustrates, neither of these assumptions is always valid.

In 1968, Ranchers Exploration and Development undertook the first commercial production of copper using the SX-EW process at its Bluebird Mine in Arizona. Since that time, hundreds of innovations have improved the process—enhancing the quality of the copper produced, reducing costs, increasing the range of treatable copper-bearing minerals, and extending the weather and other conditions for successful operation. Moreover, these developments will certainly continue into the future. This means that companies and countries that stay at the forefront of these efforts can indefinitely enjoy a cost advantage over their rivals, thanks to technology.

In addition, the SX-EW process reduces the costs of some producers much more than others. Specifically, it favors:

- Companies and countries that historically have been important copper producers. Over the years these producers have accumulated substantial waste piles of oxide copper minerals. The SX-EW process is particularly suited to recover the copper from such low-grade ores.
- Companies and countries where stringent environmental regulations are enforced. The sulfur emissions recovered from smelting copper provide a low-cost source for the diluted sulfuric acid used in the leaching step of SX-EW processing.
- Companies and countries possessing copper deposits in arid regions. The leaching step of the SX-EW process is difficult to control where precipitation is heavy.
- Companies and countries with copper reserves that contain few valuable by-products. To date, the SX-EW process has not been able to economically recover gold, silver, molybdenum, and other valuable by-products often found in copper ores.

These conditions exist particularly in the United States and Chile.[5] This explains why these two countries account for such a large share of the world's SX-EW copper production. It also explains why, in turn, the SX-EW process accounts for such a large share of their total copper output.

The SX-EW process is a particularly dramatic example of the impact on competitiveness and wealth creation that innovation and new technology can have. At the other end of the spectrum, there are thousands of small innovations that can improve the performance of individual mines. As every mine is unique, it has its own innovative opportunities. Small innovations may individually have little influence on mine productivity, costs, production, and thus comparative advantage. But when aggregated, they can be of great importance. While some of these opportunities extend over several or even many mines, many are useful only for a given mine with its unique situation. Such innovations do not diffuse rapidly, if at all, to other mines and other parts of the world.

Labor Productivity, Costs, and Mine Survival

The collapse and revival of the US copper mining industry during the 1970–1995 period raises another intriguing question: Why did some mines manage to survive and even to expand their output over this period, while others shut down?

Table 6.3 separates the 24 significant copper mines operating in the United States in 1975 into three groups. The ten expanding mines managed not only to survive, but to increase their output during the following 15 difficult years. The three contracting mines survived as significant producers but suffered a loss in output. The 11 non-surviving mines either shut down entirely or cut back to the point where they were no longer significant producers.

Economic theory and common sense leads one to expect that the expanding mines had the lowest cash costs and the highest labor productivity at the start of the period, and just the opposite to be the case for the non-surviving mines. Table 6.4 provides some support for these expectations, though there are anomalies. The non-surviving mines, for example, had lower cash costs in 1975 and higher labor productivity than the contracting mines.

Much more surprising, simple econometric models indicate that the ability of mines to reduce their cash costs and to increase their labor productivity after 1975 was far more important in explaining survival than their starting position in 1975 (Tilton, 2001). This, again, suggests that innovative activity played an important role in the recovery of the US copper industry.

Why were certain mines more successful than others in fostering productivity growth and in reducing cash costs? While many factors were likely involved, Table 6.4 indicates that the expanding mines were larger and held substantially larger reserves than the contracting and non-surviving mines at the beginning of the period. Large mines with many employees possess more human capital for innovative efforts. Given the greater number of jobs at risk, they may also be more concerned about survival. Similarly, mines with many years of reserves are likely to have greater incentives to invest in new facilities that embody the latest technology. This is because the expected returns can be realized over a longer time horizon.

Table 6.3 Output and labor productivity for 24 US copper mines, 1975 and 1990[a]

Mines	Output[b]			Productivity[c]		
	1975	1990	Growth	1975	1990[f]	Growth[f]
Expanding mines[d]						
Bagdad	20	136	590	20	102	414
Chino	53	145	172	60	91	51
Morenci	125	324	158	53	95	78
Ray	49	112	129	44	68	55
Tyrone	75	155	106	51	95	87
Bingham Canyon	247	371	50	31	153	394
Pinto Valley	66	88	34	59	77	31
San Manuel	109	142	30	22	36	63
Cyprus Miami	45	57	28	42	52	24
Sierrita	132	137	4	35	57	61
Contracting mines[d]						
Butte	91	90	−2	43	123	184
Mission[e]	106	79	−26	28	62	122
White Pine	71	51	−29	13	24	82
Non-surviving mines[d,f]						
Silver Bell	19	4	−80	45	44	−1
Mineral Park	27	2	−93	34	14	−59
Superior	44	3	−94	18	15	−18
Yerington	31	2	−94	35	30	−14
Bisbee	13	1	−96	44	35	−20
Esperanza	24	0	−99	38	49	30
Continental	16	0	−100	32	22	−32
Ajo	33	0	−100	42	29	−31
Battle Mountain	20	0	−100	30	21	−31
Ruth McGill	31	0	−100	16	14	−12
Sacaton	20	0	−100	39	45	16
All other mines[h]	75	98	30	g	g	g
Total industry[h]	1,542	1,995	29			

Sources: Brook Hunt and Associates, Rio Tinto Mine Information System, and US Mine Safety and Health Administration, as cited in Tilton (2003).

Notes:

a All US copper mines whose 1975 output equaled or exceeded 10,000 tons or more of contained copper equivalent in concentrates are included in this table with the exception of Twin Buttes. Although Twin Buttes' 1975 output was 13,800 tons of contained copper, it was excluded because its 1975 production was abnormally low, causing its productivity for that year to be unusually low as well.

b Output is measured in thousands of tons of copper equivalent contained in concentrate production. Output growth is the percent change in output between 1975 and 1990.

c Productivity is measured in tons of copper equivalent contained in concentrate produced per 1,000 man hours of labor input. Productivity growth is the percentage change in productivity between 1975 and 1990.

d Expanding mines survived the recession in the copper market during the 1975–1990 period and even managed to increase their output. Contracting mines survived the recession but suffered a decline in output. Non-surviving mines ceased to be significant producers in the sense that their output fell below 4,000 tons of copper equivalent.

e The Mission mine also includes the Pima mine.

f Labor productivity reported for non-surviving mines for 1990 is actually for their last normal year of operation: 1975 for Ruth McGill and Bisbee, 1976 for Battle Mountain, 1977 for Yerington, 1980 for Mineral Park, 1981 for Silver Bell, Superior, Esperanza, and Continental, and 1983 for Ajo and Sacaton.

g Productivity data for all other mines are not available.

h Output for all other mines is the contained copper in concentrate production, and does not include the copper equivalent of by-product output. Total industry output includes the copper equivalency of by-products for all mines except those included under all other mines.

Table 6.4 Average output, labor productivity and cost performance for expanding, contracting, and non-surviving US copper mines, 1975–1990[a]

Performance	Expanding mines	Contracting mines	Non-Survivors[d]
Output growth 1975–1990 (percent)	81	−18	−96
1975 productivity (tons/1,000 hours)	36	24	28
Productivity growth[d] 1975–1990 (percent)	125	124	−19
1975 cash costs[b] (cents per pound)	154	165	160
Cash costs growth[b,d] 1975–1990 (percent)	−42	−19	23
1975 breakeven costs[c] (cents per pound)	116	146	116
Breakeven costs growth[c,d] 1975–1990 (percent)	−29	−26	−6
1975 average output[a] (thousands of tons)	92	89	25
1975 average reserves (millions of tons)	558	126	34
1975 average reserve life (years)[e]	47	10	9

Sources: Output and productivity data: Table 6.3 and the sources cited there. Cash costs, adjusted breakeven costs, and reserves: Rio Tinto Mine Information System, as cited in Tilton (2003).

Notes:
a   See Table 6.3 for an explanation of how output and productivity are measured. This table also defines expanding, contracting, and non-surviving mines, and identifies the mines in each of these groups.
b   Cash costs are in real (1997) US cents per pound. They cover all the expenses of mining and processing through to the refined metal stage minus capital costs (specifically, depreciation, amortization, and interest on external debt). Cash costs typically include expenditures for labor, materials, energy, and contract services of third parties.
c   Breakeven costs are also in real (1997) US cents per pound. They are actually adjusted breakeven costs, which are cash costs minus any revenues received for co-products and by-products, minus the difference, if any, between a mine's reported revenues per pound of copper and the world copper price.
d   Data for 1990 reported for labor productivity, cash costs, and breakeven costs for non-surviving mines are actually for their last normal year of operation: 1975 for Ruth McGill and Bisbee, 1976 for Battle Mountain, 1977 for Yerington, 1980 for Mineral Park, 1981 for Silver Bell, Superior, Esperanza, and Continental, and 1983 for Ajo and Sacaton.
e   Reserve life for each mine is calculated by dividing the product of its reserves and the grade of its reserves by its 1975 output.

Chilean Copper Industry, 1975–2000[6]

To what extent can we generalize the finding that innovative activity is as or more important than mineral endowment in causing changes in comparative advantage in mining? There are good reasons to suspect the US situation may be an anomaly. While the country has been and remains a major copper producer, in recent years the development of most new copper mines has largely taken place abroad in Chile and elsewhere. One would expect exploration and the development of new deposits to play a more important role in these areas.

Our research on Chile was largely motivated by the desire to see if copper mining in that country enjoyed a similar jump in labor productivity during the 1980s as in the United States. And if so, to what extent innovative activity, as opposed to the development of new mines, drove the increase.

As Figure 6.3 shows, labor productivity increased in Chile during the 1980s, but at a modest pace.[7] Chile did experience a jump in productivity similar to that

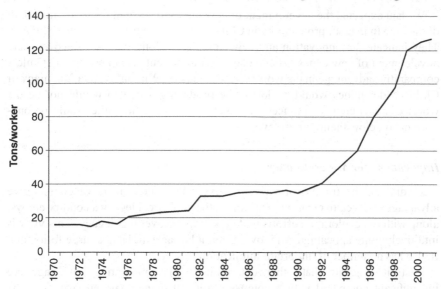

Figure 6.3 Labor productivity in the Chilean copper industry, 1970–2001 (tons of copper
contained in mine output per copper company employee) (sources: Comisión
Chilena del Cobre, Servicio Nacional de Geología y Minería, and Consejo
Minero, as cited in Tilton (2003)).

in the United States. But this occurred only in the 1990s, a decade after the surge
in the United States.

The 1990s was the decade of new mines in Chile. This suggests that better
deposits, rather than innovative activity, were largely behind the rise in labor
productivity in that country. Previously, the state mining company, Codelco,
contributed the lion's share of the country's copper output. In 1990, for example, its
mines accounted for three-quarters of all the copper mined in Chile. During the next
decade, this figure fell to nearly one-third as Escondida, Candelaria, Cerro Colorado,
Zaldivar, El Albra, Collahuasi, and other new mines came on-stream. Private
multinational mining corporations, for the most part, developed these new mines.

Our research indicates that the shift in mine output, particularly toward new
mines, accounted for about two-thirds of the jump in labor productivity during the
1990s (García *et al.*, 2001). This still leaves a surprisingly large portion of the
jump—nearly one-third—attributable to increases in labor productivity at old
mines. Chuquicamata, Salvador, El Teniente, and Andina—Codelco's traditional
mines—increased labor productivity by 37 percent, 70 percent, 70 percent and 84
percent, respectively, between 1990 and 1997. A host of different innovative
efforts largely created these impressive improvements.

Moreover, when examining labor productivity growth over a longer period—
from 1978 to 1997—we find innovative activities at the level of individual mines
to be even more important in Chile. Their contribution to the rise in labor
productivity was 45 percent, compared with 55 percent for the shift in output from
mines with low to high labor productivity (García *et al.*, 2001).

We had expected the development of new mines to account for all or almost all of the growth in labor productivity in Chile, and so were surprised by these figures. They indicate that innovation and new technology along with the discovery and development of new deposits have played an important role in enhancing Chile's comparative advantage in the world copper industry. Without innovation, many of Chile's older mines would no longer be producing, Codelco would not be the world's largest copper producer, and copper exports from Chile would be one-third or so below their current levels.

Implications for Mineral Policy

According to the traditional view, the overriding determinant of comparative advantage and wealth creation in mining is the geological legacy a country enjoys, along with the exploration efforts undertaken to uncover that legacy. This view is intuitively quite appealing. And, over time, it has accumulated a large following. It also has a number of important implications.

First, it suggests that other determinants of comparative advantage are insignificant compared with a country's mineral or geologic endowment. The generation and diffusion of new technology along with other innovations, in particular, are of secondary importance. There are two, quite different rationales for this position. The first contends that the technology of mining is mature and stagnant. The few changes that do take place do not greatly affect costs. The second recognizes that advances in technology occur. But it argues they diffuse quickly around the world, providing particular mines, companies, and countries with few opportunities to acquire a cost advantage over other producers. The first of these explanations flies in the face of considerable empirical evidence. Yet, in many circles, it is still widely accepted. The second, as we have seen in the case of SX-EW technology, does not always hold.

Second, the traditional view sees comparative advantage in mining and the wealth it creates as largely a transitory gift of nature. Companies and countries with the best deposits possess the lowest production costs and generate the most wealth. Once their deposits are exhausted, however, mine production will shift to those companies and countries with the next best set of deposits. New discoveries can also from time to time cause a change in the distribution of reserves.

Third, there is little that managers and workers can do to sustain or improve the competitiveness of any particular mine. A mine can produce only so long as it has reserves. Once these are gone, it will close. To maintain their production over the longer run, companies must replace their depleting reserves by new discoveries or by acquiring new deposits in other ways.

Fourth, the ability of governments to promote their competitiveness in mining is similarly limited. While policies that encourage domestic exploration may delay the inevitable, the depletion of the best deposits and the exploitation of the best exploration sites will eventually encourage mining companies to search abroad for new reserves. Through taxation and other means, governments can acquire some of the wealth created from their domestic mineral resources and invest it. This ensures that future generations will also benefit from the country's mineral wealth even after it is gone. What they cannot do is prevent the depletion of their mineral deposits and the loss of competitiveness that follows.

The pressing policy questions that emerge from the traditional view are:

- How long will the mineral endowment last?
- How should the wealth or rents created by mining be divided among workers, companies and their shareholders, the state as a whole, and other interested parties?
- How much of the wealth or rents should the state invest in other forms of capital to ensure that future generations continue to benefit from the country's geologic legacy after the mines are shut?

These questions lead inevitably to concerns over sustainability, intergenerational equity, and the intricacies of green accounting.[8]

However, if the traditional view is wrong or just incomplete, if innovation and new technology are important sources of comparative advantage and wealth creation in mining, as the alternative view contends, the set of important policy issues changes. The whole process becomes more internally driven. There is still wealth created and rents to be captured. But they are not predetermined gifts of nature, fixed in size, that producers—firms and countries—can effortlessly gather up. Instead, they are created by the mining companies that succeed in the global competition to reduce production costs.

Mining becomes more of a high-tech industry than commonly appreciated, where managers and workers are not helpless bystanders watching external forces unravel their predetermined fate. Instead, they are crucial players who, through their innovative efforts, influence their own destinies. While every mine eventually runs out of reserves, innovation and new technology may extend by decades the path to extinction.

The role of government shifts. Ensuring that society gets a fair share of the wealth created by mining and that this wealth is used in a manner that achieves intergenerational equity becomes less important. Instead, government should promote an economic climate that encourages and supports the innovative activities of firms and individuals. In short, public policy should focus more on how to increase the benefits flowing from mining and less on how best to divide them.

In a dynamic world with innovation and technological change, human ingenuity may keep the long-run trends in the production costs and prices of mineral commodities from rising indefinitely, a possibility we revisit in Chapter 9. This reduces concerns about sustainability and intergenerational equity.

Highlights

We are now in a position to return to the central questions this chapter explores: Why does comparative advantage in mining shift over time, and what public policies should producing countries pursue to promote their domestic mineral sectors and ensure they benefit as much as possible from their mineral resources?

Comparative advantage in mining, we have seen, shifts over time and often quite dramatically. For most mineral commodities, the top mining countries today differ from those of 50 years ago and even those of 25 years ago.

We also know that mineral endowment measured by country reserves (lagged by a decade, more or less) can explain much—though not all—of the variation in mine output among countries. Any particular country's output may be above or below the level expected on the basis of its endowment, largely as a result of its political environment and its mining history (given all the advantages of traditional mining districts).

The positive association between country reserves and mine output provides support for the factor endowment theory of international trade. Countries with abundant reserves possess a comparative advantage in mining. Indeed, this follows from the definition of reserves—the quantity of a mineral commodity found in discovered deposits that are profitable to exploit under current conditions. As a result, it is more or less a tautology and so not particularly interesting. The important question, especially given the frequent shifts in comparative advantage in mining, is what causes reserves to change over time.

The traditional and prevailing view is that changes in the geographic distribution of reserves are largely driven by the depletion of operating mines and the discovery of new deposits. The alternative view focuses on the creation of new reserves by innovation and technological change.

The copper industry in the United States, as we have seen, provides considerable support for the alternative view of comparative advantage in mining. During its dramatic turnaround in the 1980s, the United States greatly reduced its production costs. This was not by discovering new and better deposits. Instead, producers resorted to a variety of innovative activities that substantially reduced costs and more than doubled labor productivity.

In Chile, the discovery and development of new mines contributed greatly to that country's rising labor productivity, particularly during the 1990s. More surprisingly, innovation and new technology also played a critical role in sustaining Chile's comparative advantage and in contributing to the wealth created by its mining industry.

While the revival of the copper mining industry in the United States during the 1980s may be exceptional, the experience of the successful mining firms in that country and in Chile is not all that unusual. New technologies have greatly affected comparative advantage and wealth creation in the gold, nickel, and other mineral industries. Around the world, mining companies are continually searching for new technologies and other innovations to reduce costs. The discovery and development of new deposits is only one of many possible ways for countries to enhance their competitiveness in mining. Indeed, as we have seen, often innovation and new technology are as, or even more, important.

The implications for producing firms and countries that flow from the two views of comparative advantage are quite different. According to the traditional view, mineral-producing firms and governments can do little, other than support exploration, to prevent the loss over time of their competitiveness. As a result, governments should focus on capturing as much of the profits or rents from mining as possible and on ensuring that this wealth is distributed fairly between the current and future generations.

According to the alternative view, changes in competitiveness flow largely from the innovative efforts of mining companies, which as a result do have considerable influence over their future. This means a country can enhance its comparative advantage with policies that foster the innovative efforts of its domestic mining industry. This shifts the focus of public policy from ensuring the pie—that is, the wealth created by mining—is divided equitably to increasing its size.

Notes

1   Labor productivity is measured by the ratio of output to labor input—for example, the tons of nickel produced per worker in the nickel industry. Total productivity is measured by the ratio of output to all inputs (labor, capital, energy, and materials) usually weighted by their relative costs.
2   The coefficient on the X_i variable, it is worth noting, is greatly affected by the inclusion of Chile in the equation, as Chile is an outlier. If it were eliminated from the equation, the coefficient would be much smaller, perhaps half its estimated value of 0.60. This suggests that Chilean reserves in 2000 were of much better quality than the reserves of most other countries.
3   This section draws extensively on Tilton (2003).
4   This section draws heavily from Tilton and Landsberg (1999), Aydin and Tilton (2000), and Tilton (2000, 2001, 2003).
5   Today some of these conditions apply to Africa, which could over the coming decade become an important producer of copper from oxide ores.
6   This section draws extensively on García et al. (2001).
7   It is important to note that labor productivity is measured differently in Figures 6.2 and 6.3. In Figure 6.2, labor productivity is the copper contained in US mine output per 1,000 hours of work by copper company employees. In Figure 6.3, labor productivity is the copper contained in Chilean mine output per copper company employee. For an assessment of the different measures of labor productivity, within the context of the Chilean copper industry, see García et al. (2000).
8   Green accounting entails adjusting GDP and other national accounts to take account of changes in the value of a country's natural resources due to mining and other developments and also changes in environmental quality. For more on green accounting, sustainability, and intergenerational equity, see Chapter 10.

References

Aydin, H. and Tilton, J.E., 2000. Mineral endowment, labor productivity, and comparative advantage in mining. *Resource and Energy Economics*, 22, 281–293.

García, P., Knights, P.F. and Tilton, J.E., 2000. Measuring labor productivity in mining. *Minerals and Energy*, 15, 31–39.

García, P., Knights, P.F. and Tilton, J.E., 2001. Labor productivity and comparative advantage in mining: the copper industry in Chile. *Resources Policy*, 27, 97–105.

Haberler, G., 1977. Survey of circumstances affecting the location of production and international trade as analysed in the theoretical literature, in Ohlin, B., *et al.*, eds, *The International Allocation of Economic Activity*, Holmes and Meier, New York.

Tilton, J.E., 2000. Mining and public policy: an alternative view. *Natural Resources Forum*, 24, 49–52.

Tilton, J.E., 2001. Labor productivity, costs, and mine survival during a recession. *Resources Policy*, 27, 107–117.

Tilton, J.E., 2003. Creating wealth and competitiveness in mining. *Mining Engineering*, September, 15–22.

Tilton, J.E. and Landsberg, H.H., 1999. Innovation, productivity growth, and the survival of the U.S. copper industry, in Simpson, R.D., ed., *Productivity in Natural Resource Industries: Improvement through Innovation*, Resources for the Future, Washington, DC, 109–139.

United Nations, 1955. *Survey of World Iron Ore Resources*, United Nations, New York.

US Bureau of Mines, 1952a. *Materials Survey: Copper*, US Government Printing Office, Washington, DC.

US Bureau of Mines, 1952b. *Materials Survey: Nickel*, US Government Printing Office, Washington, DC.

US Bureau of Mines, 1953. *Materials Survey: Bauxite*, US Government Printing Office, Washington, DC.

US Bureau of Mines, 1976. *Mineral Facts and Problems*, US Government Printing Office, Washington, DC.

US Geological Survey, annual. *Mineral Commodity Summaries*, US Geological Survey, Washington, DC. Available at http://minerals.usgs.gov/minerals/pubs/mcs/2015/mcs2015.pdf

World Bureau of Metal Statistics, annual. *World Metal Statistics Yearbook*, World Bureau of Metal Statistics, London.

7 Market Power and Competition Policy

This chapter explores two quite different policy issues associated with the use and abuse of market power. The first, primarily a concern for consumers and consuming countries, is the fear that producers—by restricting their output—may raise market prices above a competitive level. This redistributes the rents or wealth created by mining so that producers get more and consumers less. Accentuating this fear is the fact that many important mineral producers are large, diversified, global companies operating in industries where the barriers to entry and exit are substantial.

The concern over market power and anticompetitive firm behavior, of course, is not unique to the mineral sector. It exists in many other economic sectors as well. Still, over the past century or so, competition policy—or what is called antitrust policy in the United States—has with some frequency focused on the steel, aluminum, and other mineral industries.

Competition policies are designed to prevent firms from possessing and exploiting market power. More specifically, they encourage competitive behavior by breaking up or controlling monopolies, by regulating mergers and acquisitions, and by proscribing cartels and collusion among producers.

The second policy issue concerns the potential use of market power by producing countries to enhance the benefits they and their citizens received from their mineral production. In Chile, for example, one encounters from time to time the question: Since this country accounts for one-third of global copper supply, why does it not control the market price the way Saudi Arabia and OPEC seem to control the price for oil? Would this not produce a higher market copper price and add to the rents the country receives for its valuable mineral resources?

Competition policy cannot proscribe such efforts, as governments represent independent sovereign states. Of course, Chile and other producing countries did establish a copper cartel in the late 1960s, an effort that was not a great success and ultimately was abandoned in the late 1980s (see Box 7.4). And, even for oil, just how much influence Saudi Arabia and OPEC have had in the past and now actually possess over the market price is subject to considerable debate. All of which raises the important question of when and under what conditions governments, either alone or with other governments, actually possess market power.

This chapter addresses such questions. Building on and extending the discussion of market power in Chapter 4, it first examines why economists, policy analysts, and others generally prefer competitive industries and markets. It then identifies the traditional indicators or markers of market power, including market concentration and vertical integration, and considers what these measures tell us about the presence of market power in the mineral industries. The chapter continues with an overview of competition policy, which describes how governments try to minimize the anticompetitive effects of monopolies, mergers and acquisitions, and firm collusion in the mineral sector. The final section—the Highlights section—looks at the implications of the findings for market power in the mineral sector, both the market power of producing firms and that of host governments.

Market Power and Social Welfare

When firms possess and use market power to raise prices above their competitive levels, they earn excess profits, or profits greater than those necessary to provide a competitive rate of return on their invested capital. These excess profits come at the expense of consumers.

Figure 7.1 illustrates this transfer of wealth. Figure 7.1a shows consumer surplus and producer surplus with a competitive price P*. The consumer surplus— the area under the demand curve and above the competitive price P*—reflects the difference between what consumers would be willing to pay and what they actually have to pay for the product. This is the surplus or wealth they receive from this market. The producer surplus—the area above the supply curve and under the competitive price P*—reflects the difference between the price producers actually receive and the prices at which they would be willing if necessary to sell their product. This is the surplus or wealth that this market provides to producers. Taken together, the consumer and producer surpluses constitute the social welfare or rent that the industry creates.[1]

Figure 7.1b shows how the consumer and producer surpluses change when firms raise the market price from the competitive price P* to P' by reducing their output from the competitive level Q* to Q'. As this panel illustrates, this increases the producer surplus at the expense of consumer surplus.

So the first criticism of market power is its potential to create what generally are considered arbitrary and unfair transfers of wealth. The second criticism is its potential to promote inefficiencies. Chapter 4 identified three types of efficiencies— allocative, production, and dynamic. The first concerns the level of output and whether it is above or below the optimal level. The second concerns how goods are produced. Is the optimal mix of inputs used? Are economies of scale fully realized? The third concerns the pace at which new innovations and technologies are created and adopted. Are there barriers to developing and introducing better or cheaper products?

Figure 7.1 is also helpful in illustrating the allocative inefficiency associated with the exercise of market power. When firms reduce their output from Q* to Q', the consumer and producer surpluses associated with this forgone production are

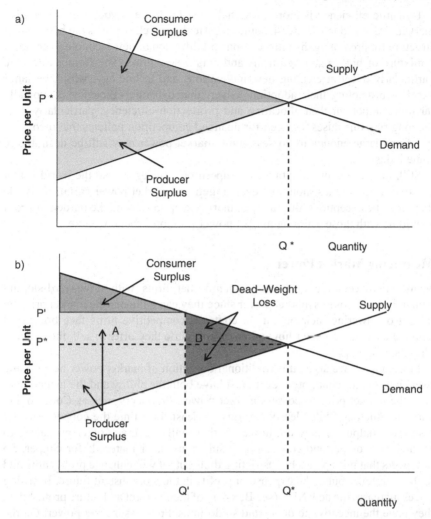

Figure 7.1 Consumer and producer surplus: (a) with competitive price; (b) with price
　　above the competitive price.

lost. This is why the fall in the consumer surplus in Figure 7.1b is greater than the
rise in producer surplus. Economists refer to this difference between what
producers gain and consumers lose as the dead-weight loss. It reflects the allocative
inefficiency that occurs when the benefits associated with the last unit produced
(measured by the market price, as it reflects the reservation price of the marginal
or last consumer) exceed the social cost (measured by the costs of the last unit
produced). In such situations, raising output by one or more units increases the
total amount of consumer and producer surplus and in turn social welfare.

　Similarly, it is fairly easy to show that the exercise of market power can cause
production inefficiency. In particular, firms restricting output to raise prices may
produce at outputs that fail to take full advantage of economies of scale.

Dynamic efficiency is more of a challenge for those advocating competitive markets. As noted in Chapter 4, some empirical research suggests the best market structure for promoting the introduction and diffusion of new technology contains a mixture of big, established firms and small, new firms. The former often are particularly adept at creating new innovations and technology, while the latter excel at promoting their adoption. Given that dynamic efficiency is typically far more important than allocative and production efficiency, particularly over the long run, this raises the need for nuanced competition policies that recognize that firms large enough to possess some market power may still be desirable in some industries.

Still, the general presumption of competition policies around the world is that competitive industries should be encouraged and market power curtailed. While there may be exceptions, they are just that—exceptions—and the burden of proof should lie with those claiming market power promotes social welfare.

Measuring Market Power

Many microeconomic textbooks contend that firms with downward-sloping demand curves possess market power since they can influence the market price by raising or lowering their output. In contrast, competitive firms face horizontal demand curves. No matter how much or how little they offer to sell, they cannot alter the market price.

In Chapter 4, we argue this traditional definition of market power needs some tweaking. In particular, producers must have both the ability and the incentive to alter the market price to possess market power. Producers such as Codelco and countries such as Chile clearly can raise at least for a time the copper price by restricting output. If they do, however, they will encourage other producers to expand their output and consumers to substitute other materials for copper. So producers that raise prices by restricting their output will enhance profits now and in the near term, but at the expense of profits in the more distant future. If raising prices increases the new NPV (see Box 4.1) of their current and future profits, then they have the incentive to do so and so do indeed possess market power. On the other hand, if raising prices decreases the NPV of current and future profits, producers do not have market power. As they have no incentive to raise prices, they will behave as competitive firms.[2]

While this is helpful in defining market power conceptually, how in practice do we know whether or not firms have the ability and the incentive to raise the market price? What measures are available to identify potentially non-competitive firms and industries? As Chapter 4 points out, a non-competitive firm is simply one with market power, and a non-competitive industry is one that has at least one non-competitive actor with the ability and incentive to influence the market price. Non-competitive actors can be producing or consuming firms with market power, producing or consuming countries with market power, cartels of producing firms or governments, as well as associations of producers and consumers (such as the former International Tin Agreement).

There are five factors widely assumed to influence market power—market concentration, vertical integration, product homogeneity, elasticity of demand, and ease of entry and exit.

Market Concentration

Market concentration reflects the number and relative market shares of producers. Although there are many ways to assess market concentration, the two most widely used measures are the concentration ratio and the Herfindahl–Hirschman Index (HHI). Fortunately, as Hall and Tideman (1967) show, there is a high correlation among the various indices. So the choice of measure is not likely to matter greatly.

The concentration ratio is simply the sum of the market shares (s_i) of the four largest firms, or the eight largest firms, or more generally of the N largest firms, as shown in Equation 7.1:

$$C(N) = \sum_{i=1}^{N} s_i \tag{7.1}$$

The global market shares of the largest four firms and the four-firm concentration ratio $C(4)$ for 24 selected mineral industries are shown in Table 7.1 for the year 2013. Market shares are reported in terms of equivalent production, rather than controlled production (see Box 7.1). While market shares based on controlled production are usually preferable, they are often unavailable, as most companies do not report their controlled production.

Table 7.1 Four-firm concentration indices for 24 mineral industries, 2013

Bauxite	Market share (%)	Lead	Market share (%)
Alcoa	18.1	Glencore Xstrata	5.5
Rio Tinto (Alcan)	16.7	BHP Billiton	4.2
US Rusal	8.6	Doe Run Company	3.5
BHP Billiton	7.4	Volcán Compañía Minera	2.6
C4	50.9	C4	15.7

Copper	Market share (%)	Manganese	Market share (%)
Freeport-McMoRan	10.2	BHP Billiton	20.0
Codelco	9.8	Eramet	8.7
Glencore Xstrata	8.2	Consmin	8.1
BHP Billiton	6.3	Vale	5.6
C4	34.5	C4	42.4

Gold	Market share (%)	Molybdenum	Market share (%)
Barrick	7.5	Freeport-McMoRan	17.4
Newmont	5.7	Codelco	8.5
Anglo American	4.3	Grupo México	7.4
Goldcorp	2.8	China Molybdenum	5.7
C4	20.2	C4	39.0

Table 7.1 continued

Palladium	Market share (%)	Tin	Market share (%)
Norilsk Nickel	36.0	Yunnan Tin Group	20.6
Anglo American	18.6	Glencore Xstrata	9.6
Impala Platinum	11.5	PT Timah	7.2
Stilwater Palladium	3.0	Minsur	7.0
C4	69.1	C4	44.3

Iron Ore	Market share (%)	Nickel	Market share (%)
Vale	22.8	Norilsk Nickel	12.8
Rio Tinto	16.8	Vale	12.0
BHP Billiton	10.7	Glencore Xstrata	10.2
Fortescue Metals Group	5.1	BHP Billiton	6.9
C4	55.5	C4	41.9

Platinum	Market share (%)	Titanium	Market share (%)
Anglo American	40.4	Rio Tinto	25.7
Impala Platinum	27.6	Tronox	12.0
Lonmin	12.1	Sierra Rutile	3.2
Norilsk Nickel	11.3	Iluka Resources	2.9
C4	91.4	C4	43.8

Rhodium	Market share (%)	Uranium	Market share (%)
Anglo American	41.6	KazAtomProm	15.8
Impala Platinum	24.9	Cameco	15.4
Lonmin	11.5	Areva	14.8
Norilsk Nickel	10.2	ARMZ – Uranium one	13.7
C4	88.1	C4	59.8

Silver	Market share (%)	Zinc	Market share (%)
Peñoles	5.7	Glencore Xstrata	10.4
KGHM	4.5	Vedanta Resources	6.3
Glencore Xstrata	4.3	Teck Cominco	4.6
BHP Billiton	4.1	Minmetals	4.3
C4	18.5	C4	25.5

Cobalt	Market share (%)	Zirconium	Market share (%)
Glencore Xstrata	20.8	Iluka Resources	26.0
Freeport-McMoRan	9.4	Rio Tinto	19.0
Gecamines	5.0	Tronox	14.0
Vale	2.9	Cristal	6.0
C4	38.2	C4	65.0

Iodine	Market share (%)	Chrome	Market share (%)
SQM	31.5	State of Kazakhstan	17.3
Cosayach	11.9	Tata Iron & Steel	7.7
ACF Minera	7.9	Glencore Xstrata	4.8
Bullmine	5.1	International Minerals Resources	2.7
C4	56.4	C4	32.5

Diamonds (2012)	Market share (%)	Lithium	Market share (%)
Alrosa	26.9	Talison	36.0
Anglo American (De Beers)	22.0	SQM	26.0
Rio Tinto	10.0	Rockwood Lithium	15.0
BHP Billiton	1.0	FMC Lithium	10.0
C4	59.9	C4	87.0

Potash	Market share (%)	Niobium (2012)	Market share (%)
PotashCorp	19.0	CBMM	80.0
Uralkali	18.0	Anglo American	6.7
Mosaic	16.0	Iamgold	6.4
Belaruskali	14.0	Grupo Parana-Panema	5.7
C4	67.0	C4	98.9

Source: our calculations based on company reports and other sources.

Box 7.1 Equivalent versus Controlled Production

Equivalent production allocates the production from a mine owned by two or more companies among its owners according to their ownership share. Controlled production, as its name indicates, allocates the production of a mine owned by two or more companies to the company that manages the mine and controls its output.

For example, consider a mine annually producing x tons of a certain mineral commodity that is 60 percent owned by Firm A and 40 percent by Firm B. The equivalent production would be $0.6 \times x$ for Firm A and $0.4 \times x$ for Firm B. If Firm A, with the largest ownership share, manages the mine, which is usually the case, then the controlled production of Firm A will be x and that of Firm B will be 0.

For a real-world example, we can consider the diamond industry. In terms of equivalent production, Alrosa, a group of Russian diamond miners, was the world's largest producer with a market share of 26.9 percent in 2013 (see Table 7.1). Anglo American had the second largest market share at 22.0 percent. However, Anglo American's De Beers subsidiary controlled about half of global production through its marketing arm, the Diamond Trading Company. Hence, Anglo American possessed by far the largest market share in terms of controlled production.

Table 7.2 shows the equivalent and controlled production for the ten largest copper-producing firms for the year 2003. Concentration ratios are higher when market shares are measured in terms of controlled production. The C(4) ratio, for example, is 36 percent with controlled production compared to only 29 percent for equivalent production.

Table 7.2 The ten largest copper producers in 2003 according to controlled and equivalent production

Company	Controlled production (%)	Company	Equivalent production (%)
Codelco	10.7	Codelco	10.9
BHP Billiton	10.3	Phelps Dodge Corp	6.5
Phelps Dodge Corp	8.8	BHP Billiton Group	6.0
Grupo Mexico	6.4	Grupo Mexico	5.5
Anglo American plc	6.1	Rio Tinto plc	5.1
Freeport-McMoRan	5.5	Anglo American plc	4.4
Antofagasta plc	3.3	Freeport-McMoRan	4.0
KGHM Polska	3.3	KGHM Polska	3.5
Xstrata plc	3.2	Norilsk Nickel	3.0
Norilsk Nickel	3.1	Antofagasta plc	2.8
C10	60.7	C10	51.7

Source: Guzmán (2006).

Interestingly, Rio Tinto enjoys the fifth largest market share with equivalent production, but only the eleventh largest (and thus does not appear in the table) with controlled production. So which production—equivalent or controlled—one uses in calculating market shares and concentration ratios can make a difference.

Four-firm concentration ratios vary greatly among the mineral commodities. Those reported in Table 7.1 range from nearly 99 percent for niobium to slightly less than 16 percent for lead, with most falling between 30 and 60 percent.

This table, along with Table 7.3, also indicates that many firms appear among the four largest in more than one industry. This may have implications for competition when assessing combined market power, since the incentives and threats to collude become larger as firms have the possibility to cooperate simultaneously across more than one market. One reason for such diversification is the similarities among mineral commodities in the technologies and expertise employed in their mining, processing, and marketing.

Another important characteristic of concentration ratios is their propensity to change over time due to the entry and exit of firms, mergers and acquisitions, and simple shifts in firm market shares. Figure 7.2 shows how the four-firm concentration ratios for 20 mineral industries changed between 1975 and 2013. A few, including iron ore, uranium, and palladium, doubled or tripled their ratios. At the other end of the spectrum, gold, molybdenum, and cobalt experienced significant declines. On balance, roughly half of the industries shown in this table saw their ratio rise while the other half saw it fall.

The HHI is the second common measure of market concentration. Though less widely used than the concentration ratio, it has the advantage that it reflects the inequality of market shares among firms. Its disadvantage is that it requires data on the market shares of all producers and so is more difficult to calculate.

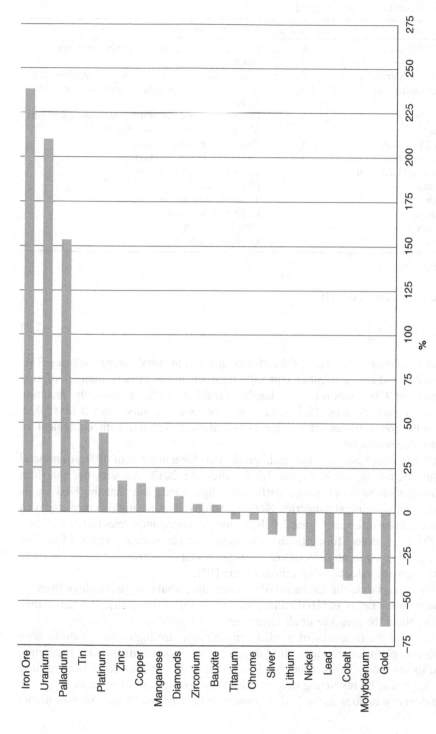

Figure 7.2 Percentage change in four-firm concentration ratios for 20 mineral commodity industries, 1975–2013.

168 *Public Policy*

Table 7.3 Mining firms with market shares among the largest four producers in two or more mineral commodities, 2013

Company	Frequency	Mineral commodities
BHP Billiton	8	Bauxite, lead, copper, manganese, iron ore, nickel, silver, diamonds
Glencore Xstrata	7	Lead, copper, tin, nickel, zinc, cobalt, chrome
Anglo American	6	Gold, palladium, platinum, rhodium, diamonds, niobium
Rio Tinto	5	Bauxite, iron ore, titanium, zirconium, diamonds
Vale	4	Manganese, iron ore, nickel, cobalt
Norilsk Nickel	4	Palladium, nickel, platinum, rhodium
Impala Platinum	3	Palladium, platinum, rhodium
Freeport-McMoRan	3	Copper, molybdenum, cobalt
Lonmin	2	Platinum, rhodium
Iluka	2	Titanium, zirconium
Codelco	2	Copper, molybdenum
Tronox	2	Titanium, zirconium
SQM	2	Lithium, iodine

Source: Table 7.1

The formula for the HHI is:

$$\text{HHI} = \sum_{i=1}^{N} s_i^2 \tag{7.2}$$

where s_i is the market share of the ith firm and N is the total number of firms. This formula produces a number that falls between 0 (an infinite number of tiny firms) and 1 (monopoly). Occasionally, especially in legal cases, the resulting figure is multiplied by 10,000, allowing the index to vary from 0 to 10,000. As with concentration ratios, the HHI can be calculated with equivalent or controlled production.

HHI indices based on equivalent production for a number of different mineral industries are shown in Figure 7.3 for the year 2013. As with the four-firm concentration ratios examined earlier, this figure indicates that the HHI varies widely across mineral industries. Not surprisingly, it is also highly correlated with concentration ratios, as Figure 7.4 shows for 23 mineral industries (all those found in Table 7.1, except for niobium). This means we can normally rely on four-firm or eight-firm concentration ratios in assessing market power and avoid the more data-intensive calculations required for the HHI.

This brings us to the fundamental question: Just what can we conclude from the concentration ratios and HHI indices for various mineral industries? What do they tell us about the presence or absence of market power?

Where these measures of market concentration are high—for example, with niobium, platinum, rhodium, lithium, palladium, potash, diamonds, zirconium, and iodine—they suggest that firms have the ability at least for a time to raise market prices by restricting output. What they do not show is whether they have the incentive to do so as well. The incentive to raise prices depends on how higher

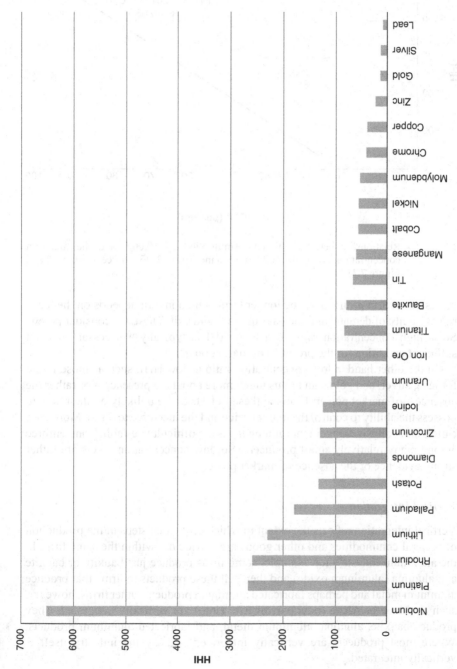

Figure 7.3 HHI for 24 mineral industries, 2013 (sources: our calculations based on company reports and other sources).

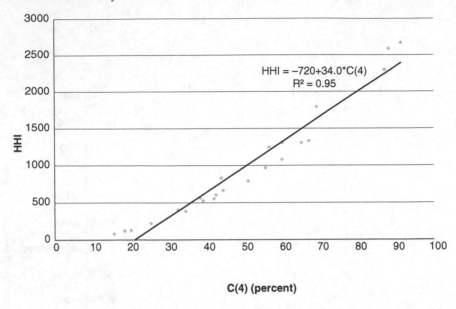

Figure 7.4 Relationship between the HHI (multiplied by 10,000) and the four-firm concentration ratio C(4) for 23 mineral industries, 2013 (sources: Table 7.2 and Figure 7.3).

prices affect firm profits over the longer term, which in turn depends on the long-run elasticity of demand and the ease of entry and exit. These we consider below. So, a high concentration ratio or a high HHI is typically a necessary but not sufficient condition for the presence of market power.

On the other hand, a low concentration ratio or low HHI, such as those found for lead, silver, and gold, can tell us much more about the presence—or rather the absence—of market power. Firms in these industries are unlikely on their own to possess the ability to control the market price and the incentive to do so. Moreover, cartels and collusive agreements among firms are difficult to establish and enforce among many relatively small producers. So, low concentration levels are rather strong evidence of the absence of market power.

Vertical Integration

Vertical integration reflects the extent to which sequential steps in the production of mineral commodities and other goods are carried out within the same firm. In the aluminum industry, for example, some firms produce just bauxite or bauxite and alumina (aluminum oxide) and then sell these products to firms that produce aluminum metal and perhaps fabricated aluminum products. Other firms, however, such as Alcoa or Alcan (now part of Rio Tinto) are vertically integrated. They produce bauxite, alumina, aluminum metal, and fabricated aluminum products. Where most producers are vertically integrated, we say the industry itself is vertically integrated.

Vertical integration raises three questions. First, has it become more or less important over time in the mineral sector? Here, as one might guess, the evidence varies from one mineral commodity to another. However, in general over the past 50 years one finds a fairly pervasive tendency for vertical integration to decline. The reasons for this are many, but the rise of first Japan and then China as major consumers of mineral commodities is particularly important. Both of these countries have strong preferences to import copper concentrates, bauxite, iron ore, and other unprocessed mineral commodities so that their own firms can carry out the downstream processing domestically.

Second, does it matter which step in the production process we use when calculating concentration ratios? We can, for example, calculate the four-firm concentration ratio for copper at the mining and milling stage, the smelter stage, or the refinery stage. Are there significant differences at each of these stages? Table 7.4 indicates that the answer is yes. It shows the market shares for the ten top producers at each of these three stages of copper production and finds the four-firm concentration ratio is appreciably higher at the mining stage. While Codelco, Freeport-McMoRan, Glencore Xstrata, and a few other firms are active at all three stages, other firms are not fully integrated. Jiangxi Copper and Aurubis, for example, are important at the smelting and refining stages but not at the mining stage.

Third, what can vertical integration tell us about the presence or absence of market power? For economists, especially those in the field of industrial organization—the economic specialty that examines in depth how market structure, firm behavior, and industry performance are related—vertical integration is an important characteristic of market structure, one that they often assume influences firm behavior and competitiveness. Just how vertical integration affects competitiveness, however, is less clear than in the case of market concentration.

In industries, such as bauxite, with both integrated and non-integrated producers, concentration ratios based on the market shares of all firms, rather than just those firms that sell bauxite, may be misleading. Whether this is actually so, however, is

Table 7.4 Market shares for the ten largest copper producers by stage of production, 2014

Rank	Mine	%	Smelter	%	Refinery	%
1	Codelco	10.0	Jiangxi Copper	6.1	Codelco	7.1
2	Freeport-McMoRan	8.0	Glencore Xstrata	5.8	Aurubis	5.1
3	Glencore Xstrata	7.2	Codelco	5.8	Freeport-McMoRan	5.1
4	BHP Billiton	6.1	Aurubis	4.3	Glencore Xstrata	4.9
5	Southern Copper	3.9	Tongling	3.7	Jiangxi Copper	4.6
6	Rio Tinto	3.0	JX Holdings	3.4	Tongling	3.4
7	Anglo American	2.9	KGHM Polska Miedz	3.1	Southern Copper	3.2
8	KGHM Polska Miedz	2.8	Sumitomo Metal Mining	3.1	JX Holdings	2.8
9	First Quantum Minerals	2.4	Freeport-McMoRan	3.0	Sumitomo Metal Mining	2.6
10	Antofagasta	2.4	Southern Copper	3.0	KGHM Polska Miedz	2.5

Source: company reports

Note: market shares are in terms of equivalent production.

not entirely certain, as firms that buy bauxite ultimately must compete with the integrated firms when they sell their aluminum metal or fabricated products. This limits what they can pay for their bauxite and, in turn, the price that bauxite producers can extract. There is still much we need to learn about what vertical integration can tell us about the prevalence of market power in mineral commodity industries.

Product Homogeneity

The automobile and many other industries sell similar products with distinctive differences in features, style, and other characteristics. It has long been known that firms producing heterogeneous products may possess some market power, some control over the price they receive, simply because consumers do not consider other products perfect substitutes for theirs.

For mineral commodities, this characteristic of market structure is often dismissed, as most mineral commodities are homogeneous. Copper or aluminum metal does not vary from one producer to another. Of course, this is not true for all mineral commodities. Ferrous scrap comes in many different grades; copper concentrates may or may not possess valuable by-products or troubling impurities such as arsenic; and gem diamonds possess so many different features that only true experts can assess their value. In addition, a ton of iron ore available in Brazil possesses a different value for a Chinese steel firm than the same ton located in Australia.

That said, it is not worth dwelling too much on product heterogeneity in the case of mineral commodities. Where differences do exist, such as those just noted, the market tends to add premiums or impose penalties, just the way it does when valuing regular and high-octane gasoline. So fluctuations in the market price for copper concentrates without arsenic are highly correlated with those with arsenic, leaving little room for producers to exploit such differences to artificially raise prices.

Demand Elasticities

Economists have long known that where the price elasticity of demand for a product is high or infinite, even a monopoly cannot restrict output and raise the market price for long. Any attempt to push the market price up will simply result in the loss of customers. Conversely, where the price elasticity of demand is low, less than 1 and close to 0, higher prices have little effect on demand.

What do we know about the price elasticity of demand for mineral commodities? This question received considerable attention in Chapter 2, where we found that the demand for most mineral commodities is indeed inelastic (less than 1) in the short run. This is in part because mineral commodities, with few exceptions, are not final goods themselves, but are used to produce final goods. They typically account for only a small fraction of the latter's production cost. A doubling of the price of aluminum or titanium, for example, increases the total costs of producing a Boeing 747 by but a small fraction. Mineral commodity prices as a result have little effect on the prices and output of many of the final goods from which their demand is derived.

In addition, the new technologies largely responsible for material substitution and resource-saving new innovations take time. As a result, they, and in turn commodity demand, are much more sensitive to changes in commodity prices over the long run than the short run.

So attempts to raise the market price by restricting output may appear successful for a while. At some point, however, they are likely to diminish demand and hurt profits. This difference between the price elasticity of mineral demand in the short and long runs is one of the principal reasons why firms with the ability to raise the market price may not have the incentive to do so and why firms—either individually or as a group—may account for a large share of the market and still not possess market power.

Ease of Entry and Exit

Low barriers to firms entering and exiting an industry can also constrain the exercise of market power. Where this is the case, a firm that attempts to artificially raise the market price stimulates the output of other producers, thereby reducing its own demand. This is the second principal reason why firms with the ability to raise the market price may not choose to do so.

Chapter 3, in looking at the nature of mineral supply, notes that there are significant barriers to entry and exit in the mineral sector. Moreover, it typically takes five years to bring new mines into operation, and under certain circumstances the time required can extend to a decade.

For this reason, once output approaches full capacity, the short-run price elasticity of supply for most mineral commodities is close to zero. Over the long run, a period of five years or so, existing and new firms can construct new capacity. Moreover, given the relative abundance of marginal deposits, the long-run supply curve tends to level off and become quite flat as output increases. As a result, the long-run price elasticity of supply, in sharp contrast to the short-run elasticity at full capacity, is quite high.

Again, this suggests that attempts to raise the market price by curtailing production may appear successful for a while. Over the long run, however, such efforts are likely to foster shrinking market shares and declining profits. When the NPV of the lost profits in the more distant future exceed the NPV of the extra profits realized over the near term, producers lack an incentive to raise the market prices and so behave as competitive firms.

To recapitulate, we have identified five common factors that economists often consider when assessing market power and industry competitiveness. Two of these—vertical integration and product heterogeneity—are of limited usefulness with mineral commodity markets. Market concentration provides useful insights into the ability of firms to raise the market price by limiting their production. The price elasticity of demand and the ease of entry and exit are particularly helpful in assessing whether firms with the ability also have the incentive to raise the market price.

Competition Policy

Public policies aimed at forbidding anticompetitive behavior date back to antiquity. In recent history, Canada passed the first competition legislation in 1889. The Sherman Act of 1890 in the United States soon followed. It provided the foundation for policies around the world. Today, most countries have competition laws and policies as well as government agencies responsible for monitoring and enforcing these laws.

Current regulations consider both industry characteristics and firm behavior. For example, the US Department of Justice and the Federal Trade Commission, in assessing mergers under the Clayton Act and monopoly under the Sherman Act, look at both the market shares of the relevant producers and the likely effects of mergers or monopolization on firm behavior. Would a merger, for example, allow the combined firms to compete more effectively with other established producers? Would it enhance the ability of the industry to innovate? In such cases, the government and the courts apply a *rule of reason*, which takes into account whether the merger or monopoly is likely to stimulate competition, promote innovation, or in other ways improve the wellbeing of society in general.

Some practices, however, are *per se violations* of the law. For example, in the United States if firms explicitly agree to set prices or divide up markets, this is all the government has to demonstrate to obtain a conviction. Firms cannot defend themselves by claiming their behavior was in the public interest and thus reasonable.

The rest of this section examines competitive policies as they apply to the mineral sector with regards to monopolies, mergers and acquisitions (M&A), and cartels and collusion.

Monopoly

Economics defines monopoly as an industry with one seller and many buyers (see Chapter 4 and Table 4.1). The legal definition of monopoly is more complicated. The process first requires that the relevant market be determined in terms of the product or products it encompasses and its geographic range. This is often both difficult and critical (see Box 2.1). The market share of the firm in question is then calculated. If it exceeds a given threshold, commonly 70 percent (though at times just 50 percent), the company is considered a monopolist.

The antitrust case that the US government brought against Alcoa following World War II for allegedly monopolizing the aluminum industry (see Box 7.2) provides a good illustration. This case also highlights some of the complications in defining the relevant market, particularly with respect to secondary production.

Box 7.2 Secondary Production and Alcoa's Market Share

Alcoa was the only significant producer of primary aluminum in North America during the first half of the twentieth century. In 1945 the US government accused the company of monopolizing the market, as its estimated share of the primary aluminum market exceeded 90 percent (Perry, 1980). Eventually Alcoa was acquitted of all charges. Nevertheless, the company was forced to spin off its Canadian operations into the independent company Alcan.

However, as several economists (Gaskins, 1972; Martin, 1982; Suslow, 1986) have noted, secondary aluminum is a near-perfect substitute for primary aluminum. If it is included in the relevant market, Alcoa's market share at the time was only 60 percent. This is still high enough to suggest possible market power, but hardly consistent with the claim of monopoly under the traditional legal definition.

After the government demonstrates that a firm's market share exceeds the accepted threshold, for a conviction it must show that the firm has somehow behaved in an anticompetitive manner. The firm on its behalf can employ the rule of reason to defend its large market share and monopoly position, arguing that its behavior has promoted, not diminished, the welfare of society. To determine whether this is or is not the case, the court reviews available documents and the testimony of knowledgeable people regarding the performance of the firm.

Even using the legal definition of monopoly rather than the more restrictive economic definition, one is hard pressed to find a monopoly among the mineral industries today. Of the 24 industries considered in Table 7.1, only in the case of niobium does the largest producer have a market share greater than 70 percent. Moreover, given the small size of the niobium market and the concentration of reserves in a few deposits, the rule of reason may well apply, as efforts to increase the number of producers would likely require the exploitation of poorer and so higher cost deposits.

Looking at producing countries rather than producing firms, some observers have accused China in recent years of monopolizing the rare earth elements and then of restricting exports to earn monopoly profits. There is no question that China had a dominant share of these markets, sufficiently large to constitute a monopoly under most competition policies. Less clear are the motives behind China's decision to restrict its exports, and the extent to which the country in fact possessed market power in this industry, given the abundance of other sources of supply around the world (see Box 7.3).

Box 7.3 China and the Rare Earth Elements

In 2010, China produced and refined around 97 percent of the global supply of rare earth elements (REEs), consisting of 17 chemical elements (the 15 lanthanides, plus scandium and yttrium). That year the Chinese government announced plans to cut exports of these elements, critical in the production of catalytic converters for cars and oil refineries, magnets, rechargeable batteries, and many medical devices. At the same time, China announced the closure of some mines and production quotas for the remaining mines.

These restrictions reduced Chinese exports from around 50,000 to 30,000 tons per year—a cut of about 16 percent in global supply. This led to a surge in prices. While many accused China of using its market power to exploit its monopoly position and earn monopoly profits, China claimed its actions were motivated by a desire to curb the overexploitation of these metals and to protect its environment.

In any case, just how much market power China has in REE markets is far from clear. The REEs, despite their name, are not particularly rare. Higher prices made the recovery of REEs from the tailings of a uranium mine in the Ukraine economic, adding some 3,000 tons to the global supply in 2011. Higher prices have also encouraged the reopening of old mines and the development of new mines in the United States, Australia, and elsewhere. Clearly any excess profits China earns from its monopoly position are evaporating along with its 97 percent market share as supply in the rest of the world expands and prices fall.

Mergers and Acquisitions

In assessing mergers and acquisitions, competition policies tend to follow similar procedures. The relevant market is defined—or markets if more than one are of concern. Then the expected market share of the merged companies following their combination is estimated. If this exceeds a specified threshold, the likely effects of the merger on firm behavior are analyzed to determine whether or not the merger will enhance competition or in other ways promote the welfare of society. The common claim of those advocating mergers and acquisitions is that they benefit society by reducing production costs.

The first document officially to set concentration limits for mergers and acquisitions was the 1982 Horizontal Merger Guidelines of the United States. Using the HHI—multiplied by 10,000, as is the custom in legal cases—this document determines two critical thresholds. The first is 1,000. Under this level, an industry is considered sufficiently atomized to allow all mergers no matter the size of the merging firms. The second threshold is 1,800. Above this level, an industry is considered sufficiently concentrated to prohibit mergers and acquisitions among the larger firms. Industries falling between the two thresholds are designated as moderately concentrated. Here, a rule of reason applies, and other factors are taken into account in judging proposed mergers. Normally, firms proposing to merge must provide evidence of synergies and lower costs.

Looking again at Figure 7.3, which shows the HHI for various mineral industries for 2013, we see that niobium, platinum, rhodium, lithium, and palladium exceed the upper threshold. Under US regulations they are considered highly concentrated industries, where mergers and acquisitions are normally not approved. Potash, diamonds, zirconium, iodine, and uranium fall between the two thresholds, and so are considered moderately concentrated. Mergers in these industries are approved or disapproved on the basis of other evidence, which weighs their costs (such as a reduction in competition) against their benefits (such as lower production costs).

For most of the mineral industries shown in Figure 7.3, however, the HHI falls below the lower threshold of 1,000. Their producing firms are sufficiently small that the approval of mergers and acquisitions can proceed without regard to their effects on industry competition. This group, it is worth noting, includes most of the more important mineral industries in terms of the value of output, including iron ore, bauxite, copper, nickel, lead, zinc, gold, and silver.

Collusion and Cartels

A cartel is an explicit collusive agreement between two or more producers whose objective is to increase the profits of its participants by reducing competition. Cartels may fix prices, restrict output (or capacity), set export quotas, establish market shares, divide markets, establish common sales agencies, and in other ways curtail competitive behavior.

There is hardly a mineral industry that at some point in history has not been cartelized. Table 7.5 identifies 23 mineral commodities where cartels have operated along with their dates of operation. The list, though not exhaustive, is sufficient to highlight the widespread incidence of cartels in the mineral sector over the past century or so.

In recent years, however, the growing adoption of competition policies has reduced the incidence of cartels among private firms. As noted earlier, in some countries explicit agreements among firms to set prices or divide markets are *per se* illegal. So the mere creation of such agreements is against the law.

However, even today some cartels may go undetected. And, of course, some cartels are legal, for example those created by sovereign states. In such cases, what can we say about their market power? What determines whether or not they can successfully raise mineral commodity prices above competitive levels or in other ways create excess profits for their members?

The answer to such questions depends on the following four factors:

1   The market share of the cartel (K). The larger the share of total production controlled by the cartel, the more likely it will be successful, everything else being equal.
2   The price elasticity of demand for the commodity (ε_D). The less market demand falls as price rises, the greater the probability of success.
3   The price elasticity of supply from outside the cartel (ε_S^{-C}). The less higher prices stimulate the entry of new firms and the expansion of existing producers outside the cartel, the better for the cartel.

Table 7.5 Mineral industries with cartels

Industry	Market	Operation years
Beryllium	World	Before World War II
Bismuth	World	Second half of nineteenth century
Bromine	Local (USA)	1885–1914
Cobalt	World	Before World War II
Copper	World	1918–1923, 1926–1932, 1935–1939, 1967–1988
Diamond	World	1934–2004
Ferrotungsten	World	1929–?
Graphite	World	1939–1940
Iodine	World	1937–?
Lead	World	1938–1939
Magnesium	World	1934–1936
Mercury	World	1928–1972
Molybdenum	World	1936–?
Phosphates	Local (USA)	1919–?
Platinum	World	World War I to ?
Potash	World	1972–current/2005–2013[a]
Radium	World	1925–?
Silver	World	1933–?
Sodium sulfate	World	1926–1939
Steel	Local (USA)	1929–1939
Tin	World	1929–?
Uranium	World	1925–?, 1972–1976
Zinc	World	1928–1929

Notes:
?   Indicates that the year the cartel ceased operating is unknown.
a   Two cartels coexisted in potash. The Eastern cartel started in 2005 and was dismantled in 2013. The Western cartel began much earlier and is still operating.

These three factors, as Barros and Vignolo (1975) demonstrate, determine the price elasticity of demand for the cartel's output ($\varepsilon_D{}^C$), as shown in Equation 7.3.

$$\varepsilon_D{}^c = \frac{\varepsilon_D + (1-K)\varepsilon_s^{-c}}{K}\tag{7.3}$$

If the price elasticity of demand for a cartel is elastic ($\varepsilon_D{}^C > 1$), a 1 percent rise in the market price causes the cartel's output to decline by more than 1 percent. This means its total revenue or sales will fall. In such cases, it is often assumed that cartels will not have an incentive to raise the market price. While this is not necessarily true—as conceivably the fall in total revenues could be more than offset by lower production costs, the result of lower output, allowing profits to rise—in practice where the price elasticity of demand for a cartel's output is greater than 1, the cartel is likely to possess little or no market power.

4 Cheating and the internal cohesion of the cartel. To the above three determinants for successful cartels, we need to add internal cohesion. To some extent, of course, cohesion depends on the preceding three determinants, but it is also influenced by other factors, such as similarities among members in culture and language, ease of detecting cheating, and the extent of differences in production costs. The stronger internal cohesion is, the less likely members are to cheat or leave the cartel.

While the extent to which these four determinants of cartel success are met varies across mineral commodities, we do know that there are big differences between the short and long runs. This is less so for the first determinant, the market share of the cartel. Whether this is large or small depends mostly on the level of market concentration and the number of large firms in the cartel.

The price elasticity of demand, however, is typically much higher in the long run than in the short run. As noted, in the long run consumers can develop the technologies that produce resource-saving innovations and that facilitate material substitution. Similarly, the price elasticity of supply from outside the cartel is much higher in the long run, as it provides the time producers outside the cartel need to construct new capacity. Internal cohesion, as well, is likely to be less favorable for the cartel over the long run. As higher prices start to erode the market share of the cartel, the incentives to cheat or to abandon the cartel tend to intensify.

So even where cartels are legal or where they go undetected, they seldom last for long. CIPEC, the copper cartel established by the governments of Chile, Peru, Zambia, and the Congo in 1967 (see Box 7.4), is a good example. Table 7.5 provides additional support for this conclusion. As it shows, the mineral industries have, over the years, seen many cartels come and go, but few have lasted for long.

Box 7.4 CIPEC

In 1967 the governments of Chile, Peru, Zambia, and Zaire (now the Democratic Republic of the Congo) created the Intergovernmental Council of Copper Exporting Countries—CIPEC, from its French name *Conseil Intergouvernemental des Pays Exportateurs de Cuivre*. CIPEC's purpose was to coordinate the policies of member countries to achieve higher copper prices and revenues for its members. Over time the governments of four other copper-producing countries—Australia, Indonesia, Papua New Guinea, and Yugoslavia—joined the organization and toward the 1980s CIPEC accounted for about 30 percent of world refined copper production and over 50 percent of proven reserves.

According to Mardones *et al.* (1984), the price elasticity of demand for CIPEC's output was in the range 1.08–2.34. Other studies obtained similar results, suggesting that this cartel possessed little or no market power. In addition, CIPEC was unable to effectively coordinate production cuts during its existence. Its internal cohesion simply was not sufficient to provide the necessary discipline and to prevent cheating. Over time its inability to exercise market power and raise copper prices became increasingly clear, leading to its termination in 1988.

Highlights

The issue of market power gives rise to two quite different policy concerns in the mineral sector. The first is the traditional fear on the part of consumers of all products (not just mineral commodities) that anticompetitive behavior provokes unearned and unfair transfers of wealth from them to producers. In addition, in the process it creates inefficiencies in production. The second is the desire on the part of producing governments to use their market power to capture as much as possible of the wealth or rents flowing from their valuable mineral resources for their own citizens.

The first of these issues has led to the creation of competition policies around the world. These policies curtail monopolies, regulate mergers and acquisition, and largely prohibit cartels among privately owned mineral-producing firms. Competition policy does not apply to sovereign states or to cartels that governments create. Nevertheless, government cartels in the mineral sector are not particularly common, and those cartels that have emerged, such as CIPEC, have not lasted for long.

The three most important indicators of potential market power for the mineral sector are market concentration, the price elasticity of demand, and the ease of entry for new firms and exit for established firms. Market concentration varies greatly: two firms largely control the world's production of niobium, while at the other extreme the top four lead producers account for less than 16 percent of that market. For most mineral commodities, concentration is below the threshold at which the US antitrust agencies consider a market even modestly concentrated. For the others, higher or moderate concentration simply suggests that the bigger firms in these industries have the ability to raise the market price by restricting their output. However, to possess market power, these firms also have to have the incentive to do so.

The incentive to raise prices depends on the other two important indicators of market power—the price elasticity of demand and the ease of entry and exit. Here, the evidence suggests that neither is favorable to the exercise of market power over the long run. In the short run, barriers to entry do exist and the price elasticity of demand is quite low. But, this is not the case in the long run. So, firms are likely to possess market power only in mineral industries where market concentration is high and where the short-run benefits of raising the market price offset the long-run costs.

Cartels can overcome some of the problems associated with low or moderate concentration if they can attract most of the larger producers. Here again, however, the four determinants that we have identified of cartel success—the market share of the cartel, the price elasticity of demand, the price elasticity of supply from outside the cartel, and the cartel's internal cohesion—suggest that even those cartels that avoid running afoul of competition policies are likely to survive for only a limited number of years.

Ultimately both the threat to consumers and the potential for producing countries associated with market power are largely constrained in the mineral sector by natural forces—specifically the intrinsic nature of mineral demand and supply over the short and long run. To many, the large, diversified, global firms populating many mineral industries appear to possess considerable market power. The same they believe holds for the governments of major producing countries. These entities, it is true, can influence the market price by curtailing production. Few, however, have the incentive to do so, as such efforts often diminish the NPV of their current and future profits. And, without an incentive to raise prices, producing firms and governments behave as competitive entities and market power does not exist.

Notes

1   A word of caution is necessary here. While economists typically measure the contribution an industry makes to the welfare of society by adding its consumer and producer surpluses, it is not correct to assume that without this industry the social welfare of the entire economy would decrease by the sum of this industry's consumer and producer surpluses. This is because, after the industry ceased to exist, consumers would increase their demand for other products, using income they previous spent on the deceased industry's product. This would shift the demand curves for these products outward and in the process increase their consumer surpluses. Similarly, firms that were formally producers in the deceased industry would redirect their labor and other inputs to other industries. This would shift their supply curves of these industries outward and change their producer surpluses. So total social welfare would fall but by less than the consumer and producer surpluses in the deceased industry, since some of this decline would be offset by increases in the consumer and producer surpluses of other industries in the economy.

2   The incentive to restrict output to raise price depends, of course, on the objective function of producers. Here we assume their objective function is to maximize the NPV of present and future profits. While this seems reasonable for most private firms, it may require some modification when assessing the incentives of producing countries to raise prices by restricting output.

References

Barros, P. and Vignolo, C., 1975. Poder monopólico en el mercado mundial del cobre. *Revista de Ingeniería de Sistemas*, 1 (1), 9–25.

Gaskins, D., 1972. Alcoa revisited: the welfare implications of the secondhand market. *Journal of Economic Theory*, 7 (3), 254–271.

Guzmán, J.I., 2006. Concentración y Competencia en las Industrias de Minerales. Mining Centre Working Paper, Catholic University of Chile, Santiago, Chile.

Hall, M. and Tideman, N., 1967. Measures of concentration. *Journal of the American Statistical Association*, 62, 162–168.

Mardones, J.L., Marshall, I. and Silva, E., 1984. *Chile y CIPEC en el mercado mundial del cobre: Frenar la producción o expandir el consumo.* Centro de Estudios Públicos (CEP), Santiago, Chile.

Martin, R., 1982. Monopoly power and the recycling of raw materials. *Journal of Industrial Economics*, 30 (4), 405–419.

Perry, M., 1980. Forward integration by Alcoa: 1888–1930. *Journal of Industrial Economics*, 29 (1), 37–53.

Suslow, V., 1986. Estimating monopoly behavior with competitive recycling: an application to Alcoa. *Rand Journal of Economics*, 17 (3), 389–403.

8 Mining and Economic Development

One of the more important policy issues puzzling mineral economists and others concerns the enigmatic relationship between mining and economic development. At first blush, most of us assume this relationship is positive. When extraction costs are less than the market price of a mineral commodity, mining creates rents. Governments can use their share of these rents to construct infrastructure, improve education, and in numerous other ways promote economic growth.

Yet, in many mining countries, per capita income has grown quite modestly or even declined over the last several decades, leading many to subscribe to what is now widely known as the *resource curse*—the view that mineral wealth and its exploitation actually impede development.

This chapter first focuses on what we can call the *traditional view*, which sees the relationship as positive, not negative. It dates back to the classical economists and, though under attack, is still quite alive. The chapter then explores the great terms of trade controversy, which Prebisch (1950) and Singer (1950) ignited over six decades ago. It challenged the traditional view and led many developing countries in the early decades after World War II to pursue autarkic policies designed to reduce their dependence on mineral production. Finally, the chapter turns to the ongoing resource curse debate, examining the empirical evidence and the conceptual arguments offered for and against the curse.[1]

The Traditional View

According to Thomas Malthus, David Ricardo, and other classical economists, mineral resources should promote economic development. The logic behind this assertion rests largely on the concept of the production function. The production function reflects the technical relationship governing how much a country can produce with given amounts of labor, capital, energy, materials, and other inputs.

They saw mineral wealth as a form of natural capital, along with agricultural land, forests, and fisheries. Natural capital in turn is part of a country's total capital, which also includes physical capital (factories, houses, roads, and other human-made structures), human capital (education, public health, and other investments in people), knowledge capital (the stock of scientific and engineering expertise), and institutional capital (the legal system and other public institutions).

Table 8.1 Mineral assets for a selected group of countries, in US dollars per capita and as a percentage of natural capital and total capital, 2000[a]

Country[b]	Value in US dollars per capita	Percent of natural capital[c]	Percent of total capital[d]
Norway	49,839	91	11
Trinidad & Tobago	30,279	98	53
Gabon	24,656	86	57
Venezuela	23,303	86	52
Canada	18,566	53	6
Russia	11,777	68	30
Algeria	11,670	88	63
Australia	11,491	48	3
Iran	11,370	81	47
Congo[e]	7,536	81	214
United States	7,106	48	1
Malaysia	6,922	76	15
Syria	6,734	77	65
Mexico	6,075	72	10
Ecuador	5,205	40	15
Chile	5,188	47	7
United Kingdom	4,739	66	1
Suriname	4,451	28	9
Denmark	4,173	36	1
New Zealand	3,596	8	1
Argentina	3,253	32	2
Colombia	3,006	46	7
Nigeria[e]	2,639	65	96
Netherlands	2,053	30	0
Brazil	1,708	25	2
Tunisia	1,610	41	4
Indonesia	1,549	45	11
Egypt	1,544	48	7
Mauritania	1,311	44	16
Romania	1,222	27	4
Guyana	1,147	11	7
South Africa	1,118	33	2

Source: World Bank (2006).

Notes:

a   Mineral assets include oil, natural gas, hard coal, lignite, bauxite, copper, gold, iron, lead, nickel, phosphate, silver, tin, and zinc.

b   The 32 countries listed in this table are those that according to the World Bank (2006) possess mineral assets per capita exceeding 1,000 dollars. The list excludes a few mineral-rich countries, such as Saudi Arabia, for which the World Bank study did not provide data.

c   Natural capital includes timber resources, non-timber forest resources, cropland, pastureland, and protected areas in addition to mineral and energy resources.

d   Total capital includes natural capital, produced capital (machinery, structures, and urban land), and intangible capital (human capital, institutional capital, and other forms of social capital).

e   For some countries with poorly educated populations and weak institutions, the World Bank estimates for intangible capital are negative. As intangible capital is part of total capital, this means that total capital for some countries is less than natural capital and in a few instances even less than mineral capital. This explains why the mineral capital of the Congo and Nigeria accounts for a larger share of total capital than of just natural capital and why the mineral capital of the Congo is more than twice its total capital.

Normally, the more capital a country possesses, the greater the output it can produce, and so the higher its per capita income or level of economic development. In the case of mineral wealth, however, there is a caveat. Deposits that lie dormant in the ground do not contribute to a country's output. To unlock their potential, society must first find and then extract mineral resources. The rise of the United States as a mineral producer helped propel that country's economic development and its rise as an industrial power in the early 1900s. Similarly well-endowed countries, such as Australia, Canada, Chile, and Russia, lagged in the exploitation of their mineral wealth and in their economic development.

Table 8.1 shows the contribution of subsoil or mineral assets (oil, natural gas, coal, as well as metals and minerals) to the natural capital and the total capital for 32 countries whose mineral wealth exceeded 1,000 dollars per capita in the year 2000. The data are from a World Bank (2006) study. Given the inherent difficulties in estimating capital assets, we should consider these calculations as rough approximations. Still, they are interesting. Mineral assets are an important part of natural capital for almost all of the countries shown. And, for a few—specifically Trinidad & Tobago, Gabon, Venezuela, Algeria, Congo, Syria, and Nigeria— mineral wealth accounts for over half of total capital.

Countries can consume or invest the rents they reap from mining. Consumption contributes to current welfare, while investment raises future welfare by enhancing the country's total stock of active capital and in turn its economic growth. This assumes, of course, the funds are invested wisely. If not, mining may provide little future benefit. In such cases, however, the problem is not mining according to the traditional view. Mining provides opportunities. If a country fails to take advantage of them, the fault lies not with mining but with the government and other entities that decide how newly created mineral capital is employed.

The Terms of Trade Debate

While the traditional view of mining and economic development has always had its critics, the first major challenge arose in the late 1940s and early 1950s with the publication of two seminal articles—one by Raul Prebisch (1950), an Argentinean economist working for the Economic Commission for Latin America in Santiago, Chile; and the other by Hans Singer (1950), a German economist with the United Nations in New York. Working independently, they focused on the terms of trade of developing countries. A country's terms of trade reflect the ratio of the prices for its exports to the prices of its imports. At the time most developing countries exported primary products and imported manufactured goods. So declining primary product prices relative to the prices of manufactured goods meant developing countries had to export greater and greater quantities of primary products for a given basket of manufactured goods.

Prebisch and Singer not only claimed that the terms of trade of primary product-producing countries had been falling, but argued that the downward trend would continue in the future for two reasons. First, they pointed out that almost all primary products were sold on competitive markets. As a result, the benefits of new cost-reducing technologies were quickly passed on to consumers in the form of lower prices. Manufactured goods, in contrast, were often sold on non-competitive markets. So managers, owners, and even employees enjoyed market power, which allowed them to keep prices from falling and thereby to capture for themselves the lion's share of the benefits of technological change.

Second, Prebisch and Singer maintained that the long-run elasticity of demand with respect to income is lower for primary products than for manufactured goods. If per capita income doubles, consumers are unlikely to double their demand for coffee, wheat, or cotton for clothing. They are more likely to use their increased purchasing power to acquire better housing and more electronic equipment. As a result, as income grows over time, the demand for manufactured goods, and in turn their prices, will rise more rapidly than the demand and prices for primary products.

As Hadass and Williamson (2002) note, the terms of trade debate raises three separate questions. First, have the terms of trade for countries exporting primary products in fact declined over the long run, as Prebisch and Singer predicted? Second, what are the principal forces or determinants behind changes in terms of trade? Third, what are the implications for public policy, particularly for developing countries that largely export primary products?

Trends

Most of the literature on the Prebisch–Singer thesis addresses the first question. This is in part because new data and new time-series estimation techniques have from time to time become available. In addition, different studies come to quite different conclusions. So we are still searching for a definitive answer to the question of whether the terms of trade of primary product producers are falling over the long run.

There is no need here to review all the relevant studies. For those interested, Hadass and Williamson (2002), Cuddington *et al.* (2007), Frankel (2010), and van der Ploeg (2011) provide good surveys of this literature.

Determinants

As noted, Prebisch and Singer predicted that the terms of trade of developing countries exporting primary products would decline over time because manufacturers, unlike producers of primary products, possess market power and so can keep prices from falling as technology reduces production costs. In addition, they argued, the lower elasticity of demand with respect to income for primary products produces smaller outward shifts in their demand curves over time than is the case for manufactured goods. Just how valid are these arguments? Or more generally, are there good conceptual reasons to expect the terms of trade of countries exporting primary products to decline persistently over time?

We know that changes over time in a country's terms of trade are due to changes in the prices of its exports and imports. These, in turn, are the result of shifts in their supply and demand curves. From Chapter 4, we also know that changes in the prices of mineral commodities over the long run, which are sold on competitive markets, are largely due to shifts in the supply curve rather than shifts in the demand curve. As explained there, this is because the long-run supply curve for most mineral commodities rises initially due to the limited number of particularly high-quality, low-cost deposits, but eventually becomes relatively flat thanks to the much greater availability of supply from more marginal deposits.

A similar argument can be made for other primary products. For example, the very best land for growing coffee or grapes for wine is scarce, while suitable land of poorer quality and hence higher costs is much more abundant.

For manufactured goods, the quality of the resources being exploited is not relevant. Unless the availability of inputs is in some other way constrained over the long run, increases in supply can occur at more or less the same per unit costs. For example, manufacturers of refrigerators or CT scanners, if given sufficient time to train new employees, to increase their capacity, and to induce their suppliers to do the same, should be able to double their output without an appreciable increase in per unit production costs. As with primary products, constant marginal costs imply a relatively flat long-run supply curve.

A flat supply curve, as we saw in Chapter 4, has important implications. In particular, shifts in the demand curve now have little or no effect on the long-run equilibrium price (see Figure 4.3a). This calls into question the second Prebisch and Singer rationale for expecting a decline in the terms of trade of developing countries exporting primary products. Over the long run, if the demand for tea and manganese grows by 0.5 percent per year and the demand for computers and recreational equipment by 6.0 percent, this differential should have little or no effect on the prices for these commodities or for the terms of trade for countries exporting tea and manganese.

What does determine changes over the long run in prices and the terms of trade are shifts upward or downward in supply curves. Such shifts occur because new technologies increase efficiency and reduce production costs. Changes in wage rates and the prices of other inputs can also be important. However, the cost-reducing effects of new technologies for most goods have offset any tendency for higher input prices to push the supply curve upward. As a result, the real prices for many primary products and manufactured goods have fallen over the long run.

There are exceptions, especially in the service sector. Education, medical care, legal services, and haircuts are all examples of labor-intensive activities where new technology has been unable to increase labor productivity sufficiently to offset rising wages. In these industries, higher costs have pushed the long-run supply curve upward, causing prices to rise. Even in these cases, however, it is changing costs and shifts in the supply curve (not the demand curve) that largely govern the long-run trends in prices.

The discussion so far has focused on the long run and on competitive markets. What can we say about the short run, continuing for the moment to focus on competitive markets? From Chapter 4, we know that for mineral commodities shifts in the demand curve greatly affect prices in the short run, at least once output approaches the capacity constraint (see Figure 4.2a). As a result, the terms of trade of mineral-exporting countries fluctuate greatly over the business cycle, rising when the global economy is booming and falling when it is in recession. We know that agricultural and other primary products also experience volatile prices over the short run, due more to droughts, pests, and other disturbances on the supply side of these markets. While this instability in primary product prices can create problems for producing countries (as we will see when we examine the resource curse), the terms of trade debate focuses on long-run trends, and the need over time for developing countries to export more and more primary products for a given basket of manufactured imports. Short-run volatility in the terms of trade of primary product exporters does not imply a long-run decline. Volatility can occur around a rising or falling secular trend.

Let us now consider the determinants of the terms of trade in the presence of non-competitive markets, as Prebisch and Singer would surely urge us to do. Here, Singer (1950, pp. 477–478) explicitly challenges the conclusion that shifts in supply curves due to changes in production costs govern long-run trends in the terms of trade of primary product producers:

> The possibility that these changing price relations simply reflect relative changes in the real costs of manufactured exports of the industrialized countries to those of food and primary materials of the underdeveloped countries can be dismissed. All the evidence is that productivity has increased if anything less fast in the production of food and raw materials, even in the industrialized countries but most certainly in the underdeveloped countries, than has productivity in the manufacturing industries of the industrialized countries....
>
> Dismissing, then, changes in productivity as a governing factor in changing terms of trade, the following explanation presents itself: the fruits of technical progress may be distributed either to producers (in the form of rising incomes) or to consumers (in the form of lower prices). In the case of manufactured commodities produced in more developed countries, the former method, i.e., distribution to producers through higher incomes, was much more important relatively to the second method, while the second method prevailed more in the case of food and raw material production in the underdeveloped countries.

This quote raises two important issues. First, is there a tendency for the markets of primary products, but not those of other traded goods, to be competitive? And second, where markets are not competitive, are prices and costs independent of each other as Singer contends? Or, do prices still move up and down with costs, indicating that shifts in supply curves influence the terms of trade for both developed and developing countries?

Both Prebisch and Singer, it is important to remember, were influenced by economic conditions in the late 1940s. At that time, the industrialized world was largely confined to North America, Europe, and Japan. Both Europe and Japan were struggling to recover from World War II. So the United States was producing most of the world's manufactured products, and many of its industries, such as the automobile and appliances industries, were concentrated and may well have possessed market power.

Since then, the recovery of Europe and then Japan, followed by the industrialization and economic development of Korea, Taiwan, Singapore, and now China and India, has completely transformed the global economy. Globalization, as both its proponents and opponents realize, is reshaping the world. Goods whose producers possess market power, who can for long keep prices above production costs and so prevent the benefits of new technology from flowing to consumers in the form of lower prices, are hard to find. Where new technologies are pushing costs down, as with computers and cell phones, prices follow. Products with stagnant or rising prices are mostly items whose costs are constant or rising. Often they are labor-intensive goods where new technology struggles to offset rising wage rates.

Moreover, even in those pockets of the economy where firms do possess market power, prices tend to follow production costs. This is the case, for example, where firms set prices on the basis of costs plus some specified markup. Pure monopolists and dominant firms with numerous small competitors will, according to microeconomic theory, expand their output up to the point where their marginal costs just equal their marginal revenue. So, when marginal costs decline, they have an incentive to reduce prices.[2]

As for producer prices for metals and other mineral commodities, Crowson (2008, p. 210) notes:

> Where producers set prices they tend to keep them fixed for long periods. They often set them not by reference to marginal costs whether their own or those of the industry, but to some form of average cost. Prices change in response to clear external stimuli, like movements in the costs of major raw materials such as crude oil, or in exchange rates.... Thus, for many years there was an observable relationship between the list prices for nickel cathode put out by the major producer, International Nickel Co (Inco), and its average costs.

So for both competitive and non-competitive markets there are good conceptual reasons to believe prices and costs move up and down together. While the empirical evidence exploring this relationship is limited—largely because many firms consider their costs proprietary and as a result do not make them available to the public—the evidence with which we are familiar for mineral commodities does indicate that prices and costs are positively correlated, moving up and down together.

For example, Figure 3.10 shows comparative cost curves for all of the world's significant copper mines at five-year intervals over the 1985–2010 period, and for 2011 as well. Costs here, it will be recalled, are the cash costs or variable costs after adjusting for by-product credits for operating mines from the lowest to the highest cost facilities. The seven curves in this figure shift to the right over time as global mine capacity grows. Of particular interest for our purposes, however, are the shifts up and down in the curve noted in Chapter 3. During the latter half of the

1980s, real copper prices rose by about 25 percent. They then declined by about 25 percent during the 1990s, leaving the real copper price in 2000 at just about its level in 1985. Since 2000, the real copper price has risen sharply by over 300 percent. Now, assuming prices and costs move together, the comparative cost curve should have shifted upward between 1985 and 1990, then downward over the 1990s, and finally up sharply after 2000. A quick look at Figure 3.10 confirms this is exactly what happened.

Figure 3.10 does not consider fixed costs, and in practice it is difficult to get reliable estimates of the fixed costs of production. Figure 8.1, however, does provide some evidence that fixed costs, like variable costs, tend to move up and down with the real price of copper. It shows the capital costs per ton of capacity for a number of major new copper mines over the 1996–2013 period. Of particular relevance for our purposes is the dramatic increase in the capital costs for those mines coming on-stream after 2006, when copper prices were rising sharply.

In addition to the empirical evidence directly linking costs and prices, numerous studies show that when mineral commodity prices rise both labor and total factor productivity fall, which in turn pushes production costs up. And, just the opposite occurs when mineral commodity prices fall. Tilton (2014) provides an extensive survey of this literature.

In summary, Prebisch and Singer would have us believe that there are two important forces governing the terms of trade of developing countries exporting primary products. The first is an imbalance in market power that causes the prices of primary products but not those of manufactured goods to fall as new technology reduces production costs. The second is the greater growth in demand that manufactured products enjoy as the global economy expands due to their higher income elasticities of demand.

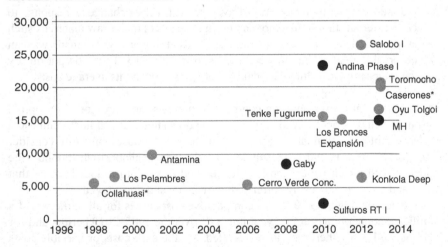

Figure 8.1 Capital expenditures in real (2011) US dollars per ton of refined copper capacity for selected new copper mines, 1996–2013 (Source: Brook Hunt and Codelco as reported in Hernández (2012, p. 16)).

Note: * Includes both concentrator and SX-EW facilities.

Upon closer inspection, however, the validity of both arguments seems questionable. With the benefit of some 60 years of hindsight, we find little evidence supporting their contention that producers of manufactured products, in contrast to producers of primary products, widely possess market power. Moreover, even where firms do possess market power, their prices tend to move up and down with production costs, indicating that even these firms share the benefits of new technology with consumers in the form of lower prices. Similarly, while manufactured products may enjoy more rapid demand growth as incomes expand, thanks to higher income elasticities of demand, this at best has little effect on the terms of trade. Product prices over the long run are largely determined by shifts in the supply curve rather than shifts in the demand curve.

What does appear important—despite Singer's protestations to the contrary—are changes in production costs. This is true for products sold on both competitive and non-competitive markets.

Since both primary products and manufactured goods benefit greatly from the cost-reducing effects of new technologies, no strong conceptual argument exists for predicting a persistent decline in the terms of trade for commodity-producing countries. This means the issue has to be resolved empirically, and as noted earlier the empirical studies are inconclusive.

Moreover, given all the unknowns surrounding the development of new technologies, even if empirical studies did consistently find primary product prices had declined relative to the prices of manufactured goods, this would not necessarily mean that they would continue to do so in the future.

Despites such uncertainties, as the next section shows, the finding that changes in production costs and shifts in supply curves largely shape long-run trends in prices and the terms of trade has important implications for public policy and for developing countries exporting primary products.

Policy Implications

Even though the debate over the actual trend in the terms of trade for developing countries exporting primary products remains unresolved, it is interesting to consider the implications of a declining trend for countries—both developed and developing—that produce and export primary products. This is particularly so since the implications are not as self-evident or as obvious as Prebisch, Singer, and their followers suggest.

Declining terms of trade mean that countries exporting primary products must, over time, offer an ever larger basket of iron ore, rice, wool, or whatever their exports are for a given basket of (non-primary) imports. This has led many to conclude that the benefits these countries reap from trade are declining. And, as a result, these countries should diversify away from primary product production.

Of course, such a strategy, if adopted by all or even many countries, would be counterproductive. As countries moved their labor and other resources out of primary production and into the manufacturing and service sectors, the prices of primary products would rise while those for manufactures and service goods would fall, reversing the decline in the terms of trade of primary products.

More importantly, economists have long known that the simple barter terms of trade—that is, the weighted prices of exports over the imports—can be misleading if used to assess the benefits of trade. As we saw in Chapter 5, the wealth or economic rent that an industry generates depends on production costs as well as the market price. And, as we have seen, prices and production costs are positively correlated, moving up and down together.

As a result, the net effect of declining prices and costs for any particular country may be positive, neutral, or negative, depending on the relative impacts of these two offsetting developments. Of particular importance for our purposes, where a country's costs are falling faster than the market price, the benefits flowing to the country from producing and exporting primary products are actually increasing despite the decline in primary product prices and the country's terms of trade. In such situations, advocating that the country introduce policies to diversify its economy away from primary products and toward manufactured and service goods seems dubious.

Conversely, where a country's costs are falling more slowly than the market price, the benefits (or rent) it enjoys from producing and exporting primary products may still be substantial, even though the benefits are declining. Moreover, the reasons they are declining is not the fall in price and the resulting deterioration in the country's terms of trade, but rather the country's failure to match the cost reductions of its competitors. This may be the result of a loss of comparative advantage in the production of primary products. If so, it is for this reason, rather than declining terms of trade, that it should be moving out of the production of primary products.

Finally, it is worth highlighting that rising export prices are not necessarily beneficial for producing countries, since rising prices normally come with higher production costs. So despite higher prices, producing countries may not enjoy greater wealth in the form of larger rents. Whether or not this is the case depends on how rapidly their costs are increasing relative to those of their competitors and in turn relative to the market price.

All of this suggests that the ongoing debate over the actual trends in terms of trade of countries producing and exporting primary products, though of intellectual interest, has limited relevance for public policy. Falling terms of trade do not necessarily mean countries should diversify away from the production of primary products. Nor do rising terms of trade necessarily mean they should encourage such production.

The failure to recognize this disconnect between the Prebisch–Singer thesis and public policy, as noted earlier, led many countries in the 1960s and 1970s to introduce policies designed to diversify their economies away from primary products by protecting inefficient and uncompetitive domestic manufacturing industries. The results for economic growth and development were sufficiently disappointing that most of these countries abandoned such policies during the 1980s and 1990s.

The Resource Curse

The first seeds of the resource curse thesis sprouted in the 1980s, ironically just about the time many countries were starting to scrap the autarky policies designed to diversify their economies away from primary products. These seeds initially were simply observations by public officials and others that many developing countries did not seem to prosper despite an abundance of mineral resources. Then a number of in-depth country studies, including those by Richard Auty (1990, 1993, 1994a, 1994b, 1994c), a British geographer who first used the term *the resource curse*, followed suggesting that mineral wealth and extraction were actually impeding economic growth.

These country studies led to more comprehensive empirical analyses that measured the effect of mining on economic development using large cross-sectional samples of developing countries. Most of these works, including the influential analyses of Sachs and Warner (1995a, 1995b, 1997a, 1997b, 1999a, 1999b, 2000, 2001), find that mineral dependence is associated with slower economic growth after controlling for the usual determinants of growth.

In this regard, Figure 8.2, based on the data that Sachs and Warner (1997a) use, is of interest. For 95 countries, it plots the average annual growth over the 1970–1990 period in real (purchasing power adjusted) GDP per person between the ages of 15 and 64 (an approximation for the economically active population) against the share of primary product exports in GNP in 1970. None of the countries with a high dependence on natural resource exports grew rapidly. Many actually suffered a reduction in per capita income. The figure clearly shows a negative relationship between per capita income growth and natural resource dependency.

Of course, a negative association between these two variables does not necessarily mean that resource dependency causes slow or negative economic growth. The causal relation between the two, some have suggested, could be the reverse. In developing countries with serious problems impeding economic growth, the resource sector, with its ability to draw on foreign capital and know-how, may be the only sector that can prosper. Alternatively, other factors adversely affecting economic growth, such as geographic location, poor institutions inherited from colonial days, or lack of access to domestic ports, may be correlated with resource dependency. However, when Sachs and Warner and others try to control for the influence of such variables, the negative relationship between growth and resource dependency persists.

Graham Davis, our colleague at the Colorado School of Mines who has devoted much of his research in recent years to the resource curse, has replicated the Sachs and Warner (1997a) analysis. Though his study (Davis, 2013) uncovers a few discrepancies, overall it confirms the validity of their results. Nevertheless, he and other skeptics of the resource curse have challenged the Sachs and Warner findings by raising questions about the time periods covered, the methodologies employed, and other aspects of these studies.[3]

Despite these reservations, the growing empirical evidence over the past several decades suggesting that mining at least in some countries is actually slowing

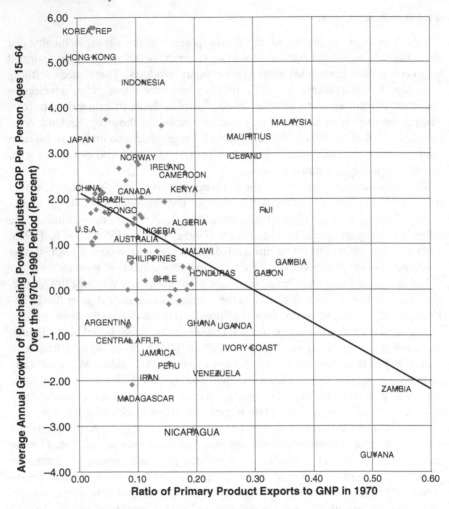

Figure 8.2 Average annual growth in real (purchasing power adjusted) GDP per person aged 15–64 over the 1970–1990 period versus the ratio of primary product exports to GDP in 1970 for 95 countries (Source: Sachs and Warner (1997a). The data are available at: www.cid.harvard.edu/ciddata/warner_files/natresf5.pdf).

economic development has led to a search for reasons why this could be so—an effort that has identified several possibilities.

The Dutch Disease

The Dutch disease is named after the adverse effects on the economy of the Netherlands produced by the expansion of that country's natural gas sector in response to the discovery of the Groningen fields in 1959. It is really a misnomer, since the Dutch disease is not a disease nor particularly Dutch.

A mineral boom of any kind requires structural adjustments within the economy, which are necessary for a country to realize the benefits from a mineral boom. Domestic wage rates throughout the economy may rise as the booming sector offers higher salaries to attract workers. As mineral exports grow, the domestic currency may appreciate against the dollar and other foreign currencies.

In assessing the effects of such developments, it is useful to divide the overall economy into three sectors—the booming mineral sector, the *tradable goods sector*, and the *non-tradable goods sector*. The tradable goods sector includes domestic industries that have to compete with foreign firms in home or overseas markets. Often these industries grow agricultural products or make manufactured goods. The non-tradable goods sector includes domestic industries that do not have to compete with foreign producers. Often they are providers of services, such as medical clinics, educational institutions, real-estate agencies, and beauty parlors.

Higher wage rates and an appreciation of the domestic currency push production costs higher. This particularly hurts the tradable goods sector, since firms in this sector cannot raise their prices without losing competitiveness to foreign producers. Firms in the non-tradable goods sector suffer less or not at all from a mineral boom, since they have more latitude to raise prices in response to higher costs.

So a mineral boom is likely to cause a contraction in the tradable goods sector. This in turn may reduce a country's economic diversity and increase its dependence on one or several mineral commodities.

It is important to stress, however, that the structural adjustments associated with the Dutch disease encourage the relocation of labor and other inputs to the expanding mineral sector. This allows the country to take advantage of its new comparative advantage in this activity. While these adjustments—like all such adjustments—create problems for certain workers and firms, particularly in the tradable goods sector, they are normally desirable and beneficial. By themselves, they are not a disease in the sense that they are something countries should try to avoid.

This means that those who claim the Dutch disease helps explain the resource curse have to go further. They have to identify a reason why the shift of domestic resources toward the booming mineral sector and hence the structural adjustments that promote this shift are actually bad rather than good for economic development.

One such possibility, which some suggest, is that once the mineral boom is over the labor and other inputs in the mineral sector will have difficulty moving back into the tradable and non-tradable goods sectors from whence they came. Why this might be so is not always clear, though economic history does suggest that such movements can be slow and difficult if workers and equipment have in the interim acquired highly specialized skills and uses.

Another possibility is that there is considerable *learning by doing* in manufacturing (see Box 8.1). So a country that reduces its manufacturing activity in response to a mineral boom may find that when the boom is over and it tries to resurrect its manufacturing sector, its costs are higher than they otherwise would have been.

Box 8.1 Learning By Doing

Learning by doing occurs when production costs fall with cumulative output. Its effects were first recognized and documented in the production of airplanes and subsequently of transistors and other semiconductor devices.

It differs from economies of scale, where production costs fall as the rate of output (for example, output per year) increases, rather than as cumulative output increases.

The benefits from learning by doing can be completely internalized, in which case cost reductions for a particular producer depend solely on its cumulative production. Alternatively, some of the benefits may be external, in which case the costs for a particular producer may depend on the cumulative production of other producers as well as its own cumulative production.

Critics of the resource curse thesis respond by noting that retraining and relocation programs can facilitate worker mobility. While capital in the mineral sector is sunk and difficult or impossible to transfer to other sectors, this is generally true of investments throughout the economy. They also note that mining and mineral processing are technologically quite sophisticated activities, where learning by doing also takes place. Moreover, history suggests that countries experiencing mineral booms normally continue to produce mineral commodities over an extended period. Indeed, it is hard to identify any such country that is no longer a significant producer of mineral commodities.

Volatile Commodity Markets

Proponents of the resource curse often point to the volatility of commodity markets in explaining why resource dependency may impede economic growth. Prices can fluctuate by 30 percent or more within a year or two. Chapter 4 highlights this volatility for metals and other mineral commodities. It also explores the underlying causes, including the substantial shifts in the demand for mineral commodities over the business cycle due to their widespread use in construction and other volatile end-use sectors.

Market instability makes it difficult for developing countries to count on revenues from the production and export of primary products. This can hamper the effective planning needed for economic development. It also means that government revenues and foreign exchange earnings are curtailed just when expansionary monetary and fiscal policies are needed to help the economy survive a downturn in a vital economic sector.

Critics of the resource curse do not deny these allegations. They do, however, tend to discount them. Downturns in commodity cycles, while they may make planning more difficult, may also promote needed changes. Government spending programs, for example, can take on lives of their own and continue long after they should be retired. With dropping government revenues, the need to cut the budget becomes clear to all, providing the rationale and political cover to cut popular but inefficient programs.

A drop in commodity prices similarly encourages producers to cut costs and improve the efficiency of their operations. As Chapter 6 notes, between 1980 and 1986, with copper prices in the doldrums, the US copper industry doubled its labor productivity and managed to survive despite widespread predictions to the contrary. During roughly the same period the iron ore industry in the Great Lakes region of Canada and the United States went through a similar metamorphosis as the rise in Brazilian imports into the region depressed prices (Schmitz, 2005). More recently, producers responded to the dramatic drop in commodity prices during the recession of 2008–2009 by increasing efficiency and reducing costs (Tilton, 2014). Many improvements introduced during such crises, such as new work rule agreements with organized labor, would be more difficult if not impossible under more normal conditions.

Moreover, even if commodity market volatility and the fluctuations in government revenues it causes do, on balance, inhibit economic growth, governments can mitigate this problem. When commodity markets are booming, they can place some of their above-normal earnings into a stabilization fund. Then, when markets fall, they can continue to support programs that they otherwise would have to curtail by withdrawing revenues from the fund.

Alaska, Canada, Chile, Ghana, Norway, Papua New Guinea, Venezuela, and other countries have created such funds. The results have been mixed, as the experience of some countries has been disappointing. Some argue that this shows stabilization funds are not a panacea. Others focus on their positive performance in other countries, and argue that better governance and stronger institutional arrangements can correct any deficiencies.

Interestingly, when Sachs and Warner (1995a, 1999a) include in their models a variable to capture the effects of commodity price volatility on economic growth, they do not find a significant link between the two. Whether this reflects the fact that fluctuations in government budgets are just not a major deterrent to economic growth or the fact that stabilization funds on balance have been successful in reducing fluctuations in government budgets is unclear.

Nature of Mining

Another explanation for the resource curse, at least in the case of mineral commodities, points to the inherent nature of mining. Mining often occurs in remote areas where it is largely an enclave activity. Needed supplies are imported, and very little added value takes place beyond mining, as ores and concentrates are shipped abroad for processing. On top of this, mining requires few workers. Those it does employ (particularly the more skilled workers) often come from abroad. As a result, it is argued, mining provides the host country with few benefits beyond the royalties and income taxes it collects.

Though not new, this line of reasoning is far from settled. Many studies show that wages and other expenditures by mining companies have an important multiplier effect on the local economy. It also is clear that in many instances mining does create downstream and upstream linkages.[4] However, the available studies focus largely on developed countries and the more advanced developing countries. It would be nice to have more evidence from the poorer developing countries, where enclave operations are more likely.

A more direct challenge to the enclave argument is simply that it is irrelevant. So what if the benefits to a country are mostly tax revenues? After all, such funds can promote education, public health, infrastructure improvements, and other activities that stimulate development.

Another troubling characteristic of mining that advocates of the resource curse often highlight concerns the distribution of the costs and benefits. Local communities often bear most of the environmental and other external costs, while the tax revenues and other benefits flow largely to the central government and the country at large. As a result, authorities representing the public interest may underestimate and undervalue the burdens borne by the local communities, and so fail to implement appropriate regulatory policies. To add insult to injury, tax revenues are often distributed in a manner that leaves nearby communities worse off, reaping more costs than benefits from mining.

On this point, local communities have, since the forced closure of the Panguna Mine in Papua New Guinea in 1989 (see Box 8.2), increasingly demonstrated an ability to prevent mine development when they believe a project is not in their best interests. Many mining companies as a result are no longer willing to risk hundreds of millions, even billions, of dollars in new projects without the support of local communities.

Box 8.2 The Panguna Mine

During the 1970s and 1980s, the Panguna Mine, located on the island of Bougainville in Papua New Guinea (PNG), was one of the world's largest open-pit operations. It produced copper, gold, and silver ores. Rio Tinto Zinc Limited owned 53.6 percent of the company that operated the mine, Bougainville Copper Limited. The PNG national government held a 19.1 percent stake in the company.

The mine was an important source of revenue for the PNG national government, accounting in some years for over 40 percent of the country's national budget. While the national government received about 20 percent of the mine's profits, the share going to the residents of Bougainville Island was only about 1 percent. Bougainvilleans were also troubled by the pollution of the Jaba River and other environmental damage. Increasingly, they felt that the company and the national government in far off Port Moresby were making the important decisions affecting their economy and quality of life with little or no input from them.

These concerns contributed to the rise of the Bougainville Revolutionary Army. The sabotage that it and affiliated forces inflicted on the mine forced its closure in 1989. To date the mine remains closed despite the occasional rumors it may soon be reopened and the fact it remains a rich mineral deposit.

Declining Terms of Trade

This explanation for the resource curse is simply a resurrection of the Prebisch–Singer thesis. From our earlier discussion of this topic, we know that the question of whether or not the terms of trade of primary product-exporting countries have declined over the long run remains unsettled. For this reason, the debate continues. We also know that the long-run trend, whatever it has been, can change. So even if the terms of trade of primary product-exporting countries have been declining, they may not continue to do so in the future.

Finally, we know that the benefits—or rents—a country derives from producing and exporting mineral commodities and other primary products depend not just on their prices but also their costs of production. Since the two tend to move together, a decline in primary product prices and the terms of trade of countries exporting these commodities does not necessarily mean the rents enjoyed by producing countries are diminishing. Moreover, even if the rents are diminishing, as long as they are positive, a case can be made for the country remaining in production.

For an entire industry, it is hard to imagine producers earning negative rents over the long run. No rational investor would build new capacity. As existing capacity came to the end of its useful life, supply would contract and eventually prices would rise, making profits and rents positive again. Of course, individual firms or individual countries may find their costs rising relative to other producers. In such cases, as noted earlier, they are losing their comparative advantage in mining and mineral production, and for this reason may need to diversify into other economic activities. But, while they are losing their comparative advantage in mineral commodities, other producers are presumably increasing theirs. So this cannot explain a general resource curse that pertains to all producers.

Use of Rents

The last explanation for the resource curse highlights the use—or misuse—of the rents derived from the production of mineral commodities. Rents flowing to the government, it is argued, mostly benefit the ruling elite. This accentuates the income disparities found between the rich and the poor, which by itself may hinder economic growth. Moreover, the mere presence of large rents may encourage individuals and organizations to devote their talents and resources to capturing a larger share for themselves. Such rent-seeking activities are unproductive; they merely redistribute the existing economic pie rather than increase it.

Even worse, as Ross (2001) and Sala-i-Martin and Subramanian (2003) have suggested, large rents may undermine the quality of a country's courts, police force, media, and other institutions. Or, as Collier and Hoeffler (1998), Gylfason (2001), and Sachs and Warner (1997a) have noted, they can lead to civil insurrection and war. Even when the rents are not squandered, but used to promote economic development, the results may be disappointing due to incompetence and poor planning.

Critics argue that good governance and strong institutions can thwart the economic incentives that give rise to rent-seeking behavior, and ensure that mineral rents are reinvested in human capital and other assets that promote growth and development. There are examples—Norway, Botswana, and Chile are often cited—that show this is true. Unfortunately, many developing countries do not have good governance and strong institutions. The pressing policy question, then, is how to ensure that the economic development of these countries is also fostered, and not retarded, by their mineral production.

Highlights

The relationship between mining and mineral wealth, on one hand, and economic growth and development, on the other, is as perplexing as it is important. The traditional view sees mining and mineral wealth as a positive force. Intuitively, this seems obvious. How could it be otherwise? How could a country with rich copper deposits, lucrative petroleum resources, or valuable gold deposits be better off without them?

And yet, many mineral-dependent developing countries have grown more slowly than countries without abundant mineral resources. Indeed, some have actually suffered a decline in their real per capita income over extended periods of time. Is this because mineral production impedes economic growth? Or is it because slow economic growth causes a country to become dependent on mineral production? Or, is it because important impediments to economic growth, not adequately controlled for in our empirical analyses, are positively associated with mineral dependence?

Although mineral economists and scholars from other disciplines are still working on these questions, we have learned a great deal over the past several decades. First, it is easy to overestimate the contribution that mineral wealth can make to economic development. The widespread presumption that countries with rich mineral resources should be rich, upon reflection, seems questionable. There is still much we do not understand about why some countries and not others—and why particular countries over some periods but not others—enjoy rapid economic growth and development. What is clear, however, is that both growth and development are complex phenomena with many interactive determinants. Among these, mineral wealth is probably a significant but relatively minor player.

That said, there is evidence suggesting that mineral production could be promoting growth and development more in a number of developing countries. Even worse, in some situations evidence suggests mineral production is actually hindering growth and development. We now have a better feel for why this may be so.

The early suggestion of Prebisch and Singer, that declining terms of trade for primary product-exporting countries is the primary culprit, is not terribly convincing. This is in part because the terms of trade of primary product producers may in fact not be declining. The bigger problem with this explanation, however, is that it ignores production costs and the rents earned from mineral production.

As long as the rents are positive, mineral production is creating wealth, wealth that has the potential to promote growth and development.

We have also seen that the Dutch disease by itself cannot explain the disappointing growth of mineral producers. For the Dutch disease to be a true disease requires the addition of fairly strong assumptions, such as the transitory nature of mineral production, sluggish factor mobility within a country, or differences in learning by doing among economic sectors. While such conditions may from time to time exist, there is little to indicate they prevail in general.

Similarly, the volatility of mineral commodity prices and the inherent nature of mining may in some situations reduce the ability of mineral production to foster growth and development. In the grand scheme of things, though, they do not seem to be of overriding importance, again because they require conditions that in the real world appear particular rather than universal. Moreover, in the case of price volatility, governments can and do create stabilization funds to dampen fluctuations in government revenues.

This leaves the use of rents as the most likely explanation as to why for some countries the contribution of mineral production to growth and development is disappointing. Where corruption, civil strife, and other rent-seeking activities squander rents, mining is not likely to contribute greatly to economic growth and may even impede it. The same is true when rents are wasted on white elephant projects or imports of expensive automobiles and luxury products. On the other hand, where the rents from mining are invested in infrastructure, education, poverty alleviation, and public institutions, they are likely to foster growth and development. Why do some producing countries spend their mineral rents wisely while others squander theirs? The short answer from much of the available research seems to be good governance and strong institutions.

What are the implications of these findings for public policy? How can mineral-producing countries and the international community promote the wise use of mineral rents and the contribution of mining to economic development?

Some argue that developing countries should just get out of the mining business. And, on a few occasions the international community seems to have agreed. It has, for example, tried to prevent the marketing of gold, diamonds, and other valuable minerals in countries, such as the Congo, when these products were funding rebel groups and promoting armed conflict. This drastic policy response, however, seems appropriate only for rather rare and extreme situations. If applied broadly it would deny developing countries access to one of the few sources of wealth they possess and can draw on to grow and develop.

Mineral wealth is an asset—a form of capital—that can help and has helped countries develop. So the important question is not whether public policy should encourage or discourage mining in developing countries. But rather, how can policy promote the good governance, strong institutions, and other favorable factors that ensure countries realize the full contribution mining can make to their economic growth and prosperity.

Notes

1   This chapter draws heavily on Davis and Tilton (2005) and Tilton (2013).
2   As Scherer and Ross (1990, p. 200) point out, it can be shown in the Cournot model
     of oligopoly prices, where firms assume their competitors' output is fixed, that the
     ratio of the difference between the market price and the weighted average industry
     marginal costs to market price equals the ratio of market concentration (measured by
     the Herfindahl–Hirshman index) to the elasticity of market demand. As a result, a
     decline in marginal costs causes the market price to fall assuming the market
     concentration and the elasticity of demand remain the same.
3   See, for example, Davis (2011), James (2015), Lederman and Maloney (2007), Smith
     (2015), and Wright and Czelusta (2004).
4   See, for example, Aroca (2001), Clements and Johnson (2000), Stilwell *et al.* (2000),
     and de Ferranti *et al.* (2002).

References

Aroca, P., 2001. Impacts and development in local economies based on mining: the case of
     the Chilean II region. *Regional Policy*, 27 (2), 119–134.
Auty, R.M., 1990. *Resource-based Industrialization: Sowing the Oil in Eight Developing
     Countries*, Clarendon Press, Oxford.
Auty, R.M., 1993. *Sustaining Development in Mineral Economies: The Resource Curse
     Thesis*, Routledge, London.
Auty, R.M., 1994a. Industrial policy reform in six large newly industrialized countries: the
     resource curse thesis. *World Development*, 12, 11–26.
Auty, R.M., 1994b. *Patterns of Development: Resources, Policy, and Economic Growth*,
     Edward Arnold, London.
Auty, R.M., 1994c. The resource curse thesis: minerals in Bolivian development, 1970–90.
     Singapore Journal of Tropical Geography, 15 (2), 95–111.
Clements, K.W. and Johnson, P., 2000. The minerals industry and employment in Western
     Australia: assessing its impacts in federal electorates. *Resources Policy*, 26 (2), 77–89.
Collier, P. and Hoeffler, A., 1998. On the economic causes of civil war. *Oxford Economic
     Papers*, 50 (4), 563–573.
Crowson, P., 2008. *Mining Unearth*, Aspermont UK, London.
Cuddington, J.T., Ludema, R. and Jayasuriya, S.A., 2007. Prebisch–Singer redux, in
     Lederman, D. and Maloney, W.F., eds., *Natural Resources: Neither Curse Nor Destiny*,
     Stanford University Press, Palo Alto, CA, and World Bank, Washington, DC,
     103–140.
Davis, G.A., 2011. The resource drag. *International Economics and Economic Policy*, 8,
     155–176.
Davis, G.A., 2013. Replicating Sachs and Warner's working papers on the resource curse.
     Journal of Development Studies, 49 (12), 1615–1630.
Davis, G.A. and Tilton, J.E., 2005. The resource curse. *Natural Resources Forum*, 29,
     233–242.
de Ferranti, D., Perry, G.E., Lederman, D. and Maloney, W.F., 2002. *From Natural
     Resources to the Knowledge Economy: Trade and Job Quality*, World Bank,
     Washington, DC.
Frankel, J.A., 2010. *The Natural Resource Curse: A Survey*, Natural Bureau of Economic
     Research Working Paper 15836, Cambridge, MA.

Gylfason, T., 2001. *Natural Resources and Economic Growth: What is the Connection?*, CESifo Working Paper No. 530, Munich, Germany.

Hadass, Y.S. and Williamson, J.G., 2002. *Terms of Trade Shocks and Economic Performance 1870–1940, Prebisch and Singer Revisited*, working paper, Harvard University, Cambridge, MA. An earlier version of this paper is available from www. nber.org/papers/W8188 as an NBER working paper.

Hernández, D., 2012. Desafíos y Oportunidades de la Minería en América Latina, presentation to Expomin, Codelco, Santiago, Chile.

James, A., 2015. The resource curse: a statistical mirage? *Journal of Development Economics*, 114, 55–63.

Lederman, D. and Maloney, M.F., 2007. *Natural Resources: Neither Curse Nor Destiny*, Stanford University Press, Palo Alto, CA and World Bank, Washington, DC.

Prebisch, R., 1950. *The Economic Development of Latin American and its Principal Problems*, United Nations, Economic and Social Council, Economic Commission for Latin America, reprinted in *Economic Bulletin for Latin America*, 1961, 7 (1), 1–22.

Ross, M.L., 2001. Does oil hinder democracy? *World Politics*, 53, 325–361.

Sachs, J.D. and Warner, A., 1995a. *Natural Resource Abundance and Economic Growth*, National Bureau of Economic Research Working Paper No. 5398, Cambridge, MA.

Sachs, J.D. and Warner, A., 1995b. *Economic Reform and the Process of Global Integration*, National Bureau of Economic Research Working Paper No. 5039, Cambridge, MA. Reprinted in *Brookings Papers on Economic Activity*, 95 (1), 1–95, 108–118.

Sachs, J.D. and Warner, A., 1997a. *Natural Resource Abundance and Economic Growth*, HIID Working Paper, Harvard University, Cambridge, MA. Available at www.cid. harvard.edu/ciddata/warner_files/natresf5.pdf

Sachs, J.D. and Warner, A., 1997b. Sources of slow growth in African economies. *Journal of African Economies*, 6 (3), 335–376.

Sachs, J.D. and Warner, A., 1999a. The big push, natural resource booms and growth. *Journal of Development Economics*, 59, 43–76.

Sachs, J.D. and Warner, A., 1999b. Natural resource intensity and economic growth, in Mayer, J., Chambers, B., and Farooq, A., eds., *Development Policies in Natural Resource Economies*, Edward Elgar, Cheltenham, 13–38.

Sachs, J.D. and Warner, A., 2000. Globalization and international competitiveness: some broad lessons of the past decade, in Porter, M.E., *et al.*, eds., *The Global Competitiveness Report 2000*, Oxford University Press, New York.

Sachs, J.D. and Warner, A., 2001. Natural resources and economic development: the curse of natural resources. *European Economic Review*, 45, 827–838.

Sala-i-Martin, X. and Subramanian, A., 2003. *Addressing the Natural Resource Curse: An Illustration from Nigeria*, National Bureau of Economic Research Working Paper No. 9804, Cambridge, MA.

Scherer, F.M. and Ross, D., 1990. *Industrial Market Structure and Economic Performance*, third edition, Houghton Mifflin, Boston, MA.

Schmitz, J.A., Jr., 2005. Lessons from the dramatic recovery of the U.S. and Canadian iron ore industries following their early 1980s crisis. *Journal of Political Economy*, 113 (3), 582–625.

Singer, H.W., 1950. The distribution of gains between investing and borrowing countries. *American Economic Review*, 40 (2), 472–485.

Smith, B., 2015. The resource curse exorcised: evidence from a panel of countries. *Journal of Development Economics*, 116, 57–73.

Stilwell, L.C., Minnitt, R.C.A., Monson, T.D. and Kuhn, G., 2000. An input–output analysis of the impact of mining on the South African Economy. *Resources Policy*, 26 (1), 17–30.

Tilton, J.E., 2013. The terms of trade debate and the policy implications for primary product producers. *Resources Policy*, 38, 196–203.

Tilton, J.E., 2014. Cyclical and secular determinants of productivity in the copper, aluminum, iron ore, and coal industries. *Mineral Economics*, 27, 1–19.

van der Ploeg, F., 2011. Natural resources: curse or blessing? *Journal of Economic Literature*, 49 (2), 366–420.

World Bank, 2006. *Where is the Wealth of Nations? Measuring Capital for the 21st Century*, World Bank, Washington, DC.

Wright, G. and Czelusta, J., 2004. The myth of the resource curse. *Challenge*, 47 (2), 6–38.

9   Depletion and Scarcity

The topic of the previous chapter—the role of mining in economic development—is of special interest to mineral-producing countries. This chapter turns to a subject—mineral depletion and scarcity—of particular concern to consuming countries.

Before delving into this issue, however, we need to define some terms, starting with *scarcity*, *shortage*, and *availability*. Scarcity and shortage, as used here, are synonyms. Where there is a shortage there is scarcity, and vice versa. Availability in contrast is just the absence of scarcity or a shortage.

When economists talk about shortages, they often are thinking of situations where the available supply is not sufficient to meet existing demand. In market economies, such situations are rare, since price normally rises, reducing demand and increasing supply, until the two are in equilibrium. So shortages can occur only if for some reason prices are blocked from rising sufficiently to balance supply and demand. In Chapter 4 we saw this can occur during market booms when firms sell their output at a producer price (see Figure 4.2b). It also occurs when governments impose price controls.

For our purposes, however, this common definition of a shortage is too narrow. If sharply higher prices are needed to bring supply and demand into balance, consumers will find the commodity increasingly difficult to afford. For them, it is in short supply and scarce. So here the terms *shortage* and *scarcity* include situations where real prices are rising sharply in the short run or persistently over the long run, in addition to those situations where demand exceeds supply.

We also distinguish between mineral commodities, such as refined copper, and *mineral resources*, such as chalcopyrite and other copper-containing minerals, from which copper is extracted. Mineral resources are the legacy of geologic processes that occurred over geologic time, measured in hundreds of millions of years. Because the time required for their formation is so vast from the perspective of humans, we consider mineral resources to be *non-renewable*. Other resources, such as water, air, forests, fish, and solar energy, are *renewable*, as they can be replaced within a relatively short period of time. So current use need not result in less availability in the future. One should not, however, make too much of this difference as both renewable and non-renewable resources can experience scarcity and both can be depleted.

Shortages can arise for a host of different reasons, which we can separate into two groups. The first contains just one threat—mineral depletion. The second contains all the other threats, such as wars, embargoes, cartels, export restrictions, strikes, mine accidents, global economic booms, and insufficient investment in new capacity. Shortages associated with this second group of threats are temporary. They tend to erupt suddenly and come as a surprise. Though they often persist for only a few months and rarely beyond a decade or so, they can be quite costly and disruptive while they last.

Countries have adopted various policies to mitigate or avoid such shortages. Following World War II, for example, the United States and the Soviet Union subsidized domestic mining to reduce dependence on less secure imports. Normally, however, the most efficient and cost-effective defense against this type of shortage is stockpiling. Inventories, whether held by governments or by firms, can replace interrupted supplies and, if necessary, provide that crucial window of time needed to develop alternative sources of supplies.

Given their temporary nature, wars, embargoes, and the other threats in the second group do not pose a serious long-run challenge for modern civilization, with its diverse and extensive material requirements. This is fortunate as they occur with some frequency. For example, the failure of the global mining industry to foresee the surge in Chinese demand during the first decade of the twenty-first century led to underinvestment in new capacity and in turn sharply higher mineral commodity prices for a number of years. Another recent example is the surge in prices of rare earth minerals in 2011, following the decision of China—at the time the source of about 97 percent of the world's supply—to curb its exports.

In almost all respects, shortages arising from mineral depletion, the sole perpetrator of scarcity in our first group of threats, are different. These shortages would emerge slowly over decades, perhaps centuries. Once in place, they would likely persist for a long time and perhaps become permanent. As a result, they could pose a serious long-run threat to modern civilization. Stockpiling is not an effective antidote. So it is fortunate that shortages due to resource depletion have been rare, so rare in fact that it is hard to identify any global examples. This, of course, does not mean that they will not arise in the future.

The rest of this chapter focuses in more depth on depletion and the threat it poses.[1] The next section provides a brief overview of the concerns found in the literature and describes the ongoing debate over this threat to humanity. Following sections then explore the two common paradigms used to assess depletion, the past impact of mineral depletion, and its possible future impact. The chapter concludes by highlighting the major findings and some of the more interesting implications that flow from these findings.

Evolving Concerns

Concern over depletion and availability of natural resources dates back over two centuries to the classical economists. Thomas Malthus, whose *An Essay on the Principle of Population* first appeared in 1798, is well known for his dismal prediction that population growth coupled with the limits on agricultural land would cause the human condition to sink to the subsistence level. David Ricardo extended Malthus' analysis by recognizing different qualities of agricultural land. The best land is used first. As population grows, land of poorer quality is brought into cultivation, pushing food prices higher and allowing owners of the more futile land to earn economic rents. These, as we saw in Chapter 5, economists often refer to as Ricardian rents.

The next wave of interest in depletion and resource availability surfaced toward the end of the nineteenth century, with the Conservation Movement. The closing of the US frontier and the cutting of vast tracts of virgin forests help spawn this largely social and political movement. Its leaders, unlike Malthus and Ricardo, were not economists. Some, such as Theodore Roosevelt and Gifford Pinchot, were public figures. Many others were natural scientists. As a result, the Conservation Movement had no coherent economic core. Declining physical supplies directly equated with a reduction in resource availability, as the following frequently cited quote from *The Fight for Conservation* (Pinchot, 1910, pp. 123–124) illustrates:

> The five indispensably essential materials in our civilization are wood, water, coal, iron, and agricultural products.... We have timber for less than thirty years at the present rate of cutting. The figures indicate that our demands upon the forest have increased twice as fast as our population. We have anthracite coal for but fifty years, and bituminous coal for less than two hundred. Our supplies of iron ore, mineral oil, and natural gas are being rapidly depleted, and many of the great fields are already exhausted. Mineral resources such as these when once gone are gone forever.

The Conservation Movement was largely concentrated in North America during the period 1890 to 1920, though similar concerns arose in other industrialized countries and at other times. For example, W. Stanley Jevons (1865) warned Britain that its limited coal resources threatened its future industrial growth.

During the 1930s the world suffered through the Great Depression. With the collapse of commodity prices, concerns over resource availability evaporated. They returned towards the end of this decade and throughout the first half of the 1940s, but the focus was largely on the short-run shortages associated with World War II. Shortly after the war, though, worries over the long-run availability of resources resurfaced as the world contemplated the reconstruction of Europe, Japan, and the Soviet Union, along with the needs of developing countries.

In the United States, these concerns prompted President Harry Truman to create the President's Materials Policy Commission, more commonly known as the Paley Commission after its chair, William S. Paley. In its five-volume report published in 1952, it offered the following assessment of the world's mineral resources (President's Materials Policy Commission, 1952, p. 2):

The nature of the problem can perhaps be successfully over-simplified by saying that the consumption of almost all materials is expanding at compound rates and is thus pressing harder and harder against resources which, whatever else they may be doing, are not similarly expanding. This Materials Problem is thus not the sort of "shortage" problem, local and transient, which in the past has found its solution in price changes, which have brought supply and demand back into balance. The terms of the Materials Problem we face today are larger and more pervasive.

Such views led to many studies over the subsequent decades on the long-run availability of mineral and other resources. One of the most influential was *Scarcity and Growth* by Barnett and Morse (1963). It drew a sharp distinction between the physical availability and economic scarcity and in turn between *physical depletion* and *economic depletion*. Whale oil provides a good example of these differences. During the latter half of the nineteenth century, the supply of whale oil for lamps and lighting collapsed as many species were hunted almost to extinction. The arrival of low-cost petroleum products and electricity, however, prevented the physical decline of whale oil from producing economic scarcity.

Using economic measures of scarcity, Barnett and Morse found that renewable and especially non-renewable resources had become more, not less, available between 1870 and 1957, the period they examined, notwithstanding the explosion in consumption taking place over these years. They credit this favorable outcome largely to new technology and its ability to offset the negative impact of resource depletion. At the time this finding was quite surprising and stood in stark contrast to the perceived wisdom of the day. It also led Barnett and Morse to conclude that future mineral scarcity due to depletion was not inevitable.

Another very influential study, which we encountered in Chapter 5, was the Hotelling (1931) analysis of exhaustible resources. Though published before World War II, it was during the second half of the twentieth century that this contribution was widely recognized. In addition to defining user costs or what many now call Hotelling rents (see Figure 5.2), this seminal work showed that under certain conditions[2] the value of mineral resources in the ground will increase at a constant rate r, where r is the expected rate of return on alternative investments with similar risks.

While Hotelling focused just on the behavior of a single mine, others have extended his analysis to cover the world as a whole. If the value of mineral resources in the ground rises persistently at some constant rate r, as Hotelling contends, eventually the dwindling stock of remaining resources will result in growing scarcity. However, this conclusion depends on his assumptions. Once we allow for exploration and the discovery of new resources or for technological change and lower costs of finding and producing mineral resources, shortages due to depletion are no longer inevitable.

The most recent wave of concern over the long-run availability of mineral commodities emerged in the early 1970s with the publication of the widely read book, *Limits to Growth* (Meadows *et al.*, 1972). On the basis of a sophisticated computer simulation model, it concludes that the world as we know it may collapse as a result of resource depletion by the middle of this century. Lending credence to this dire prediction, many commodity prices surged upward shortly after the book's appearance, the result of a global economic boom and in the case of oil the policies introduced by OPEC (the Organization of Petroleum Exporting Countries).

While *Limits to Growth* caused considerable consternation about resource availability and the human condition, many took issue with both its methodology and conclusions. The lively debate that ensued continues to this day. On one side are pessimists, such as Ayres (1993), Kesler (1994), and Deffeyes (2001, 2005). Often geologists, engineers, and other scientists, they see depletion as a serious future threat. On the other side are the optimists, such as Simon (1981), Lomborg (2001), and Beckerman (2003). Often economists (despite the discipline's reputation as the dismal science), they see little or no threat from depletion even over the long run.

During the 1990s as the prices of many commodities fell, the nature of the debate shifted somewhat. Pessimists focused more on the environmental and other social costs associated with mineral resources. Even if physical availability were not an issue, they argued, scarcity could arise for environmental reasons. We revisit this possibility in Chapter 10.

Two Paradigms

Most of us use one of two mental models to assess the threat of depletion. The first is the *fixed-stock paradigm*, the second the *opportunity-cost paradigm*.

The Fixed-Stock Paradigm

This logical construct starts with the self-evident observation that the earth is finite. This means the supply of nickel or any other mineral commodity must also be finite and thus a fixed stock. Demand, on the other hand, is a flow variable. It continues year after year. So it is only a question of time before demand consumes the available supply. Moreover, if demand is growing exponentially, as has been the case at times for many mineral commodities, the exhaustion of the available supply is likely to come sooner rather than later due to the tyranny of exponential growth.

This view of depletion is found in *Limits to Growth*, and in many other studies, including Hubbert (1962, 1969), Campbell (1997), and Deffeyes (2001, 2005). The human race is in a way like a colony of mice with the good fortune to find a huge warehouse full of cheese. Today the mice are fat, happy, and multiplying. But, the day must come when the last bit of cheese is gone, the warehouse is bare, and starvation looms.

Under the fixed-stock paradigm, astute observers may see the remaining stocks declining, but the actual transition to scarcity is abrupt and sudden, like a car running out of gas on an autobahn. Given society's myopia, the end is likely to come as a surprise. Mineral consumption accelerates the day of reckoning. The more we consume today, the sooner we run out. Population growth and rising per capita income accelerate the depletion process, while conservation, recycling, and the substitution of renewable for non-renewable resources buys time and puts off the day of reckoning.

With the fixed-stock paradigm one can calculate the life expectancies for various mineral commodities. This requires estimates of the available stock and its use over time.

Possible candidates for the available stock are reserves, resources, and the resource base. *Reserves*, it will be recalled, are the quantities of a mineral commodity found in deposits that are both known (that is, discovered) and profitable to exploit given existing technology, prices, and other current conditions. *Resources* include current reserves plus potentially feasible reserves. Potentially feasible reserves take into account deposits that are economic but not yet discovered, that are discovered but not yet economic, and that are both undiscovered and currently uneconomic, but which in all cases may with some likelihood become reserves in the future. This is admittedly a rather nebulous definition, though one that the US Geological Survey and other organizations find sufficiently concrete for estimation.[3] Finally, the *resource base* encompasses resources plus all the remaining quantities of a mineral commodity found in the earth's crust.[4]

Figure 9.1 shows the relationship between these three possible measures. Reserves are located in the upper left-hand corner of the diagram, where economic feasibility and geologic assurance are the highest. As one moves diagonally down and to the right of the diagram, profitability and certainty decline. The figure is not drawn to scale, as the resource base is many orders of magnitude greater than reserves and even resources. The McKelvey box[5] and other similar diagrams used

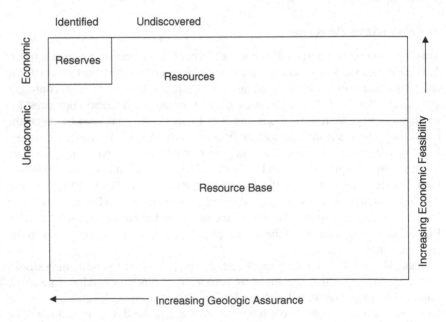

Figure 9.1 Reserves, resources, and the resource base.

by the US Geological Survey and others do not include the resource base. This is because most of the resource base is unlikely to ever become economic and so is of little or no practical interest. For our purposes, however, the minerals in the resource base are part of the available physical stock even though they may not be potentially profitable.

Table 9.1 shows the life expectancies using reserves as the available stock for oil, copper, and other selected mineral commodities, assuming their primary production grows at 0, 2, and 5 percent per year in the future. This range of growth rates seems reasonable since it brackets the average annual growth of primary production over the past several decades for all the commodities considered. The reported estimated life expectancies vary from a high of 109 years for coal to a low of 11 years for lead, which is not terribly comforting. However, we know that reserves are not really fixed stocks. New discoveries and new technologies can and do add to reserves over time, undermining the importance we can attach to these estimated life expectancies.

At the other extreme, Table 9.2 uses as its measure of the available stock the resource base, which counts all of a mineral commodity found in the earth's crust. One might question whether reserves and the resource base are necessarily the

Table 9.1 Life expectancies of world reserves, selected mineral commodities

Mineral commodity[a]	2015 reserves[b]	2012–2014 average annual primary production[b]	Life expectancy in years, at three growth rates for primary production[c]			Average annual growth in production, 1975–2014[d] (percent)
			0%	*2%*	*5%*	
Coal	8.9×10^{11}	8.2×10^9	109	57	37	2.3
Crude oil	1.7×10^{12}	3.2×10^{10}	53	36	26	1.2
Natural gas	1.9×10^{14}	3.4×10^{12}	55	36	26	2.7
Aluminum	2.8×10^{10}	2.6×10^8	108	57	37	2.9
Copper	7.0×10^8	1.8×10^7	39	28	21	2.6
Iron	1.9×10^{11}	3.1×10^9	62	40	28	3.3
Lead	8.7×10^7	5.4×10^6	16	13	11	1.2
Nickel	8.1×10^7	2.4×10^6	34	25	19	2.8
Silver	5.3×10^5	2.6×10^4	20	16	13	2.6
Tin	4.8×10^6	2.8×10^5	17	14	12	0.7
Zinc	2.3×10^8	1.3×10^7	17	14	12	2.5

Sources: US Geological Survey (2015a, 2015b); BP (2015).

Notes:

a   For the metals other than aluminum and iron, reserves and primary production are measured in terms of metal content. For aluminum and iron, reserves and primary production are measured in tons of ore.

b   Reserves and primary production are measured in metric tons except for crude oil, measured in barrels, and natural gas, measured in cubic meters. Reserves are measured at the beginning of the year.

c   Life expectancy figures were calculated before reserve and average production data were rounded. As a result, the life expectancies shown in columns 4, 5, and 6 may deviate slightly from the life expectancies derived from the reserve data shown in column 2 and the annual primary production data shown in column 3.

d   The average annual growth rate for coal shown in column 7 is for the period 1981 to 2014, rather than the period 1975 to 2014.

Table 9.2 Life expectancies of the resource base, selected mineral commodities

Mineral commodity	Resource base[b] (metric tons)	2012–2014 average annual primary production[c]	Life expectancy in years, at three growth rates in primary production			Average annual growth in production, 1975–2014[d] (percent)
			0%	2%	5%	
Coal[a]	n.a.	8.2×10^9	n.a.	n.a.	n.a.	2.3
Crude oil[a]	n.a.	3.2×10^{10}	n.a.	n.a.	n.a.	1.2
Natural gas[a]	n.a.	3.4×10^{12}	n.a.	n.a.	n.a.	2.7
Aluminum	2.0×10^{18}	2.6×10^8	7.7×10^9	951	404	2.9
Copper	1.5×10^{15}	1.8×10^7	8.4×10^7	723	311	2.6
Iron	1.4×10^{18}	3.1×10^9	4.5×10^8	808	346	3.3
Lead	2.9×10^{14}	5.4×10^6	5.4×10^7	701	303	1.2
Nickel	2.1×10^{12}	2.4×10^6	8.7×10^5	492	218	2.8
Silver	1.8×10^{12}	2.6×10^4	7.0×10^7	713	308	2.6
Tin	4.1×10^{13}	2.8×10^5	1.5×10^8	752	323	0.7
Zinc	2.2×10^{15}	1.3×10^7	1.6×10^8	757	325	2.5

Sources: the data on the resource base are derived from information in Erickson (1973, pp. 22–23) and Lee and Yao (1970, pp. 778–786). The figures for the 2012–2014 average annual production and the annual percentage growth in production for 1975–2014 are from Table 9.1 and the sources cited there.

Notes:
a   Estimates of the resource base for coal, crude oil, and natural gas do not exist. As a result, data for the resource base and life expectancies for these commodities are not available (n.a.). The US Geological Survey and other organizations do provide assessments of ultimate recoverable resources for oil, natural gas, and coal. While these are at times referred to as estimates of the resource base, they do not attempt to measure all the coal, oil, and natural gas in the earth's crust. As a result, they are more appropriately considered as resource estimates, rather than assessments of the resource base.
b   The resource base for a mineral commodity is calculated by multiplying its elemental abundance measured in grams per metric ton times the total weight (24×10^{18}) in metric tons of the earth's crust. It reflects the quantity of that material found in the earth's crust.
c   Primary production is measured in metric tons except for crude oil, measured in barrels, and natural gas, measured in cubic meters.
d   The average annual growth rate for coal shown in column 7 is for the period 1981 to 2014, rather than the period 1975 to 2014.

two extreme estimates of the available stock, since some reserves may ultimately prove unexploitable due to future government regulations or other developments, and since the resource base excludes possible future production from below the earth's crust, from the oceans, and from outer space. Nonetheless, it seems reasonable to assume that the available stock for a mineral commodity is greater than its reserves and less than its resource base.

The life expectancies shown in Table 9.2 again assume that primary production grows at an annual rate of 0, 2, and 5 percent. At current rates of production, they indicate that the resource base would last for 870,000 years for nickel to 7.7 billion years for aluminum. These large numbers suggest the human race probably has other, more pressing policy issues. However, if we assume primary production grows at 2 or 5 percent per year, these huge figures fall to mere hundreds of years, illustrating the tyranny of exponential growth and the impossibility of anything growing at a constant exponential rate for long.

If reserve life expectancies are too low, those for the resource base are presumably too high, since the world is unlikely to chew up the entire earth's crust in its quest for mineral commodities. This leads some to argue that resources are our best estimate of the available stock. But resources, like reserves, are not really fixed stocks. They change over time as the definition of potentially feasible reserves expands. We have seen this with oil over the past half-century, as oil sands, shale oil, and other unconventional sources of supply have become viable. In the case of copper, the US Geological Survey estimated copper resources at 1.6 billion tons in the early 1970s. It then added an additional 0.7 billion tons for copper contained in seabed nodules in the early 1980s. More recently, it has increased its estimate of land-based copper resources substantially, so that total copper resources now stand at over 3.7 billion tons, more than double its estimate just 45 years ago (Tilton and Lagos, 2007; Edelstein, copper specialist, US Geological Survey, personal communication). So resources, though they may be better than reserves, are still far from an ideal measure of the available stock.

At a more fundamental level, however, the fixed-stock paradigm suffers from several fatal flaws. First, many mineral commodities, especially the metals, are not destroyed when consumed. So recycling and reuse are possible. Of course, recycling in some cases, such as the lead once used as an additive in gasoline, is prohibitively expensive, but this is an issue of costs and not of physical availability. The stock of lead existing in and on the earth is the same today as it was centuries or millennia ago, aside perhaps from the negligible amounts shot into space.

Second, with few exceptions, society does not need natural resources as such, but rather the functions they fulfill. For many mineral commodities, and in particular the energy minerals, substitution may mitigate the adverse effects of depletion. Coal, natural gas, petroleum, nuclear, hydropower, geothermal, wind, and solar energy can all generate electric power. The mix of these resources that society uses at any time largely reflects their costs.

In light of such substitution opportunities, the depletion of a particular resource poses a problem only if all the alternatives are similarly suffering from growing scarcity. While the resource base for many of the non-renewable energy resources is unknown, the availability of renewable energy sources, particularly solar power, is for all practical purposes unlimited (see Box 9.1).

Third, the physical quantities of many mineral commodities found in the earth's crust are huge. For example, as Table 9.2 indicates, the quantities of copper and aluminum would last 84 million years and 7.7 billion years at current production rates. These are big numbers. For comparison, earth itself is only about 5.5 billion years old, and *Homo sapiens* as a species has existed for only several hundred thousand years.

Fourth, long before the last ton of copper or aluminum were extracted from the earth's crust, costs would rise, at first curtailing but eventually eliminating demand. In short, what we have to fear is not physical depletion, but economic depletion, where the costs of producing and using materials rise to the point where they are no longer affordable. This brings us to the second perspective or model of depletion.

Box 9.1 The Availability of Solar Power

The availability of solar power reaching earth's upper atmosphere equals the solar constant (SC) times the area of earth presented to the sun. The SC is the rate of arrival of energy per unit area perpendicular to the sun's rays at earth's location. This equals 1,350 watts per square meter. The area of earth presented to the sun equals πR^2, where R is the radius of earth (6.38×10^6 meters) and π is the well-known ratio of the circumference of a circle to its diameter (3.14159). So the solar power reaching the upper atmosphere is $SC\pi R^2 = 1350 \times 3.14159 \times (6.38 \times 10^6)^2 = 1.73 \times 10^{17}$ watts. Since only about 50 percent of this energy reaches the ground, the total solar power reaching earth's surface is half of this figure, or 8.6×10^{16} watts. Multiplying this figure by the number of hours in a year (24×365) and then dividing by 1,000 (to convert from watts to kilowatts) indicates that 7.9×10^{17} kilowatt-hours of solar energy reach earth's surface annually.

To comprehend the magnitude of this figure, we can compare it to the energy derived annually from global petroleum production. The amount of energy in a barrel of oil varies. For the United States it averages about 5.8 million BTU, or the equivalent of 1.7 thousand kilowatt-hours. Annual global crude oil production averaged 2.4×10^{10} barrels over the 1997–1999 period. At 1.7 thousand kilowatt-hours per barrel, this output contains 4.0×10^{13} kilowatt-hours of energy, or approximately 0.005 percent of the solar power reaching earth's surface every year.

According to the US Energy Information Administration, crude oil production accounts for 40 percent of global energy output. So total energy output currently equals 0.012 percent of available solar power. This means that the physical availability of solar power is some 8,000 times greater than the combined total of the world's current energy production.

The point, it is important to stress, is not to suggest that some day the world may use all of this available solar power. The costs, including the environmental costs, of solar power presumably would rise sufficiently to make the additional use of solar power uneconomic long before the globe was completely smothered by solar panels. The point is simply that it is costs, and not physical availability, that ultimately determines the availability of energy commodities.

Sources: Tilton (2003, p. 25) and the sources cited there.

The Opportunity-Cost Paradigm

This approach to assessing depletion is quite different. It focuses on what society has to give up in order to obtain another ton of copper or aluminum and how this sacrifice is changing over time. Three different measures are available for this purpose: (1) trends in the real price of the mineral commodity; (2) trends in the real extraction costs for marginal producers; and (3) trends in the value of marginal reserves.[6] By far the most widely used measure is real price, as price data are much more readily available and reliable than is the case for extraction costs and the value of marginal reserves.

Real price trends, while far superior to physical measures of scarcity (reserves, resources, and the resource base), do have their own shortcomings and limitations.[7]

For example, as noted earlier, scarcity and shortages can arise for reasons other than resource depletion. As a result, and as Chapter 4 highlights, mineral commodity prices fluctuate greatly over the short run. For this reason, long-run price trends provide the most useful insights regarding mineral depletion.

In addition, prices reflect only those environmental and other social costs that producers and ultimately consumers actually pay. This means that prices underestimate the full opportunity costs to society of mineral commodities. Of greater importance for our purposes, this also means that trends in real prices will overestimate the rise in scarcity if government policies are forcing producers to internalize an increasing share of the total social costs and underestimate the rise in scarcity if the opposite is the case. We return to this issue in Chapter 10, which addresses the environmental costs associated with producing and using mineral commodities in more detail.

Perhaps the greatest shortcoming of real prices arises when we look to the future. The only prices we have are historical, which largely reflect current and past conditions. Of course, when the market anticipates future scarcities, current prices will rise as consumers and others build up their inventories. Still, given the long-run nature of the depletion threat, it seems doubtful that current and past prices are particularly useful in forecasting availability decades or centuries in the future.

Still, how we consider depletion matters. With the fixed-stock paradigm, scarcity is inevitable. It is just a matter of time before demand consumes the available stock. With the opportunity-cost paradigm, mineral commodities may become more or less available, depending on whether the cost-reducing effects of new technology are greater or smaller than the cost-increasing effects of depletion. So it is fortunate that the opportunity-cost paradigm is the better of the two methods for assessing depletion.

The Past

A review of the many available studies of historical trends in real commodity prices leads to two general conclusions.[8] First, over the past century or two, despite the extraordinary surge in global consumption, depletion has not created serious shortages. Consumption of nearly all mineral commodities today is as great as it has ever been. The long-run trends for some commodity prices, such as aluminum, have been declining. For others, they have been more or less constant. What we do not see is rising real prices significantly curtailing consumption.

Second, the historical evidence also documents that the long-run trends in real prices and in turn the availability of mineral commodities are not fixed. Rather, they have varied over time with changes in the pace that new technologies are introduced, in the rate of world economic growth, and in the other underlying determinants of commodity supply and demand. This finding cautions against using past trends to predict future prices.

The following quote from Neumayer (2000, p. 309) nicely captures both of these lessons from the past:

So far, the pessimists have been wrong in their predictions. But one thing is also clear: to conclude that there is no reason whatsoever to worry is tantamount to committing the same mistake the pessimists are often guilty of—that is the mistake of extrapolating past trends. The future is something inherently uncertain and it is humans' curse (or relief, if you like) not to know with certainty what the future will bring. The past can be a bad guide into the future when circumstances are changing. That the alarmists have regularly and mistakenly cried 'wolf!' does not *a priori* imply that the woods are safe.

The Future

We have seen that the fixed-stock paradigm, despite its intuitive appeal, suffers from several fatal shortcomings. In addition, while historical prices are useful for documenting past trends, they are unlikely to provide reliable forecasts of future resource availability. How, then, should we proceed? What, if anything, can we say about the future?

The Cumulative Availability Curve

An alternative approach, more likely to produce useful insights than extrapolations of past price trends, uses the *cumulative availability curve*[9] to assess the important underlying determinants of long-run mineral supply and demand. The cumulative availability curve, illustrated in Figure 9.2, shows the amount of a mineral commodity that can be recovered profitably at various prices from different types of mineral deposits *over all time*. As the copper price rises from 2.00 to 5.00 dollars per pound, for example, poorer quality and hence more costly deposits can be economically exploited. Thus, the curve rises monotonically with price, as all three of the curves in Figure 9.2 show.

The cumulative availability curve is quite different from the traditional supply curve found in introductory economic textbooks. The latter depicts how the supply of a good varies with price over a year or some other given time period. Here, supply is a flow variable that can continue indefinitely. Cumulative availability, on the other hand, is a stock variable. It shows how much is available over all time. Another important difference is that the traditional supply curve indicates what will actually be offered to the market rather than what is recoverable (in the sense that it can be profitably produced) at various prices.

The concept of cumulative availability makes sense only for non-renewable resources. For renewable resources, such as fish or timber, there is no limit on cumulative availability as long as production during any particular period does not exceed the level that allows the resource to replenish itself.

Like the traditional supply curve, the cumulative availability curve assumes that, other than price, all the determinants of availability (such as the costs of labor and other inputs) remain fixed at current or given levels. Exploration, new discoveries, and the development of new mines can occur, but technology, including exploration technology, remains unchanged.

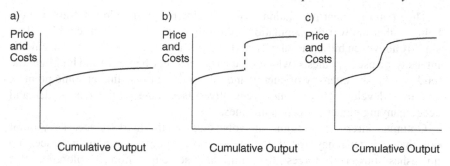

Figure 9.2 Illustrative cumulative availability curves. (a) Slowly rising slope due to gradual increase in costs; (b) discontinuity in slope due to jump in costs; (c) Sharply rising slope due to rapid increase in costs
Source: Tilton and Skinner (1987).

The cumulative availability curve is a useful expository device because it allows us to separate all the different factors affecting mineral commodity prices into three groups or categories. In the first are those variables that determine the shape of the curve. Here we find the various geologic factors affecting future costs, such as the nature and incidence of mineral occurrences. The second group contains the variables that govern primary production and how quickly the world moves up the curve. Here are all the determinants of current and future mineral commodity demand, including per capita income growth, population growth, and changes in consumer preferences. Government policies and other factors influencing recycling are also in this group because the greater secondary production is, the less primary production is required to satisfy any given level of demand. In the third group are all the forces that can cause shifts in the cumulative availability curve. Changes in wage rates and the prices for other inputs belong to this group. Over the long run, though, the most important variables in this group are typically innovation and technological change. They shift the curve downward by reducing production costs and increasing the quantities that can be profitably produced at any given price.

The first two groups reflect the cost-increasing effects of depletion, the last the cost-reducing effects of new technology. Mineral commodities will become more available in the future if, and only if, the tendency for mineral prices to rise as society moves up the curve is more than offset by downward shifts in the curve. Many are optimistic that this will be the case. Many others are skeptical and more pessimistic.

Differing Perspectives

The optimists, though well aware that past trends need not continue indefinitely, note that technological change has over many decades successfully offset the cost-increasing effects of mineral depletion. Moreover, this success has taken place during a period of exploding mineral resource use due to population growth and rising per capita incomes in many parts of the world.

They point out that population growth is slowing.[10] The United Nations now believes that the world's population, currently over seven billion, could peak at slightly above ten billion around 2100 (United Nations, 2013).[11] They also cite the intensity-of-use hypothesis (which we encountered in Chapter 2) and highlight the tendency for the intensity of energy and material use to decline at some point as countries develop. This should eventually offset some of the rise in demand accompanying higher per capita incomes.

Optimists stress, in addition, the robustness of the marketplace. They point out that any tendency for depletion to drive mineral commodity prices up unleashes powerful forces that mitigate scarcity—more exploration for undiscovered and lower-grade deposits, more development of alternative sources of supply, more substitution from scarce to abundant resources, more recycling, and more conservation.

The pessimists have far less confidence in the marketplace. They believe it irresponsible to assume that new technology will forever offset the cost-increasing effects of depletion. In the shorter run, they worry about the possibility of a 50 percent or so increase in world population over this century, and surging mineral demand in China and elsewhere in the developing world.

Assessing Cumulative Availability

Whether the future conforms more closely to the predilections of the optimists or the pessimists depends on: (a) the shape of the cumulative availability curve; (b) the speed at which society moves up the curve; and (c) the extent to which the curve shifts over time. With a fair amount of certainty (which is rare when talking about the more distant future), we know there is little hope of reliably predicting either of the last two developments. The extent to which the cumulative availability curve will shift depends on the introduction and diffusion of new technologies. Both are notoriously difficult to predict over the near term, let alone over the next century. Similarly, how rapidly society will move up the curve depends on changing consumer preferences for mineral-intensive products, population growth, per capita income growth, and trends in recycling, all of which are similarly difficult to anticipate over the longer term.

The shape of the cumulative availability curve—the first consideration—is more tractable. It depends on the nature and incidence of existing mineral occurrences. It is true that our current knowledge of sub-economic resources for many mineral commodities is quite limited. This is largely because exploration focuses on finding economic and particularly highly economic deposits. Nevertheless, it is clearly possible to obtain information on sub-economic mineral occurrences. These resources were created in the past, in many cases hundreds of millions of years ago, and ignorance about them largely reflects a lack of interest rather than intractable difficulties in obtaining the needed information.

Moreover, the shape of the cumulative availability curve can provide useful insights about the potential threat of depletion. This is true even though reliable information regarding how fast society will move up the curve and to what extent the curve will shift over time is unknown and probably unknowable. To highlight this possibility, consider the cumulative availability curve in Figure 9.2a, whose upward slope is quite gentle. As cumulative production proceeds, the price needed to elicit additional quantities increases but at a modest and decreasing rate. If this is the true shape of the curve, then new technology should find it relatively easy over time to offset the cost-increasing effects of depletion, lending support to the optimists. However, for at least two reasons—joint production and the Skinner thesis—the true cumulative supply curve may for certain mineral commodities be far less benevolent than the one shown in Figure 9.2a.

Joint Production

Cobalt, gallium, indium, lithium, the platinum-group metals, thorium, and many other mineral commodities are produced as by-products and co-products. As a result, a substantial share of their total mining and processing costs are borne by their main or co-products. Should their demand at some point exceed the supply available from by-product or co-product output, these commodities would have to be produced as main products. This could require a substantial increase in price to cover higher production costs, causing a jump or break in their cumulative availability curves. In such situations, the cumulative availability curve is likely to look more like those in Figures 9.2b and 9.2c.

The Skinner Thesis

Brian Skinner is a well-known professor of economic geology at Yale University. He points out that 99 percent of the weight of the earth's crust is accounted for by nine common elements—oxygen, silicon, aluminum, iron, calcium, magnesium, sodium, potassium, and titanium. While the cumulative availability curves for the mineral commodities in this group may follow the benign pattern shown in Figure 9.2a, he believes that the curves for copper, lead, tin, zinc, and the other commodities not in this group may possess sharply rising slopes or even discontinuities, such as those Figures 9.2b and 9.2c portray. For some mineral commodities, the curves could even have multiple jumps or steps.

According to Skinner (Skinner, 1976; Gordon *et al.*, 1987), these jumps occur because the geochemical processes creating mineral deposits millions of years ago probably did not produce a unimodal relationship (as is often assumed) between the grade and available quantity of a mineral commodity, such as that shown in Figure 9.3a. This relationship, instead, is likely to possess two or more peaks, as shown in Figure 9.3b. This means that once society has exhausted the higher-grade deposits, it will have to exploit much lower-grade, and so much more costly, resources.

In addition, the processing methods required to liberate copper and other metals in very low-grade deposits may differ from those in use today. In particular, mechanical and chemical processes for concentrating the ore before treatment may not be feasible. In this case, the energy required could be one or two orders of magnitude greater, causing a sharp jump in costs. Using copper as an example, Skinner shows in Figure 9.4 how energy requirements could increase if the world had to extract copper from low-grade silicate rock rather than high-grade sulfide ores.

The Skinner thesis is not universally accepted, even among geologists. Moreover, as Skinner himself points out, it is based on theoretical analysis. Very little empirical work has been carried out on processing very low-grade deposits, largely because they are currently of no commercial interest. As a result, sharp jumps and discontinuities in the cumulative supply curve, though possible, are not certain. In addition, if jumps and discontinuities do exist, we have little or no idea when society will encounter them, how large they will be, or how serious a threat they will pose.

As a result, for copper and most other mineral commodities, the information needed to estimate the cumulative availability curve is not currently available. There are, though, a few interesting exceptions. Aguilera *et al.* (2009) provides cumulative availability curves for petroleum products; Yaksic and Tilton (2009) for lithium; and Jordon *et al.* (2015) for thorium. Ideally, we would like such estimates to reflect availability from both known (discovered) and unknown mineral deposits. Aguilera *et al.* actually attempt to do this. In the cases of lithium and thorium, however, estimates of unknown resources are not available. So the curves estimated for these two commodities, it is important to note, are likely to shift downward over time as a result of new discoveries as well as new technology. Figure 9.5 shows the cumulative availability curve for lithium found in Yaksic and Tilton (2009). Lithium is of particular interest these days as lithium batteries are expected to power the next generation of hybrid and electric automobiles. This could cause a huge surge in lithium demand, leading many to ponder whether this means sharply higher prices and shortages.

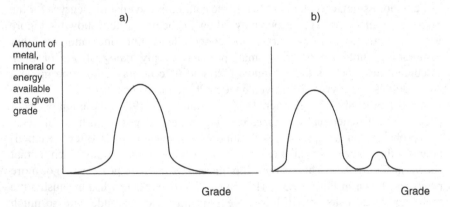

Figure 9.3 Two possible relationships between ore grade and the metal, mineral, or energy content of the resource base: (a) unimodal; (b) bimodal

Source: Skinner (1976).

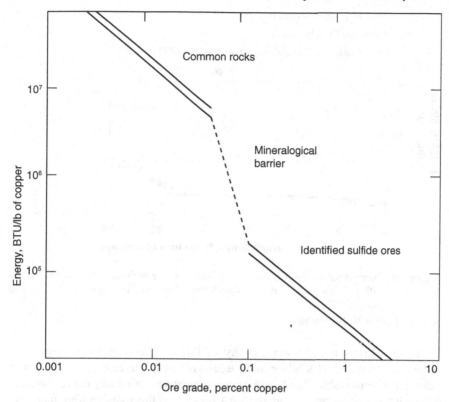

Figure 9.4 Energy required per pound of copper from sulfide ore and common silicate
rock

Source: Skinner (1976).

Andrés Yaksic constructed Figure 9.5 for his MS thesis (Yaksic, 2008). With
information from industry and government officials and from the available
literature, he identified known lithium resources and estimated their quantities and
production costs. Estimating production costs is challenging. This information,
when it exists, is often considered propriety. And, for resources not yet being
exploited, production costs do not exist and can only be approximated. For this
reason, Figure 9.5 shows costs under two scenarios—a high-cost scenario (the top
curve) and a low-cost scenario (the bottom curve). Costs are shown on the vertical
axis in dollars per pound of lithium carbonate, the most important lithium
chemical. The horizontal axis reports availability in terms of tons of contained
lithium. There is one ton of lithium in 5.323 tons of lithium carbonate.

It is important to note that the curves shown in Figure 9.5 are incomplete.
Lithium can be produced from seawater at an estimated cost of 7–10 dollars per
pound of lithium carbonate. Only a small segment of the cumulative availability
curves associated with lithium production from seawater (that is, the horizontal
segment at ten dollars for the top curve and at seven dollars for the bottom curve)

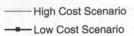

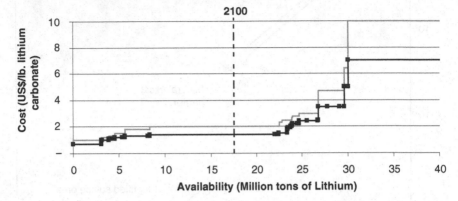

Figure 9.5 Cumulative availability curves for lithium under high- and low-cost scenarios with predicted cumulative demand from 2008 to 2100 (source: Yaksic and Tilton (2009)).

Note: Costs are in 2007 US dollars.

is shown. This is because the quantity of lithium available from seawater is huge—an estimated 44.8 billion tons assuming just 20 percent of the lithium in seawater is recoverable. The length of the cumulative availability curve shown in Figure 9.5 would increase by more than 1,000-fold if this tonnage were included.

The broken vertical line drawn in Figure 9.5 at 17.5 million tons indicates the cumulative primary production of lithium required over the rest of this century to meet the very optimistic projection of demand growth made by Yaksic and Tilton (2009). Since they expect actual demand to be less, this is a worst-case demand forecast for availability. It suggests that at some point in the twenty-second century, the world could find it attractive to extract lithium from seawater. This, according to Figure 9.5, would require a price of 7–10 dollars per pound of lithium carbonate. While ten dollars per pound represents a tripling or so over the current price, it would not greatly increase the cost of producing lithium automobile batteries or in turn the cost of electric cars.

Two caveats are worth noting, as they reinforce the conclusion that depletion does not pose a major threat to the long-run availability of lithium. First, the cumulative availability curves shown in Figure 9.5 are based on conservative assumptions and estimates. The assumed recovery rates are low. In addition, the curves exclude various known but uneconomic sources of lithium as data on their availability or costs are not available. Second, and even more important, new discoveries and new technologies will certainly shift the curves shown down by the twenty-second century.

Highlights

Shortages of mineral commodities arise quite frequently for a host of different reasons. They often occur suddenly and unexpectedly. Rarely do they persist for longer than a few years, and so do not pose a long-run threat. Nevertheless, while they last they can be disruptive. Fortunately, private and public stockpiles offer an effective and efficient mechanism for coping with these common scarcities.

The exception to the above are shortages arising from mineral depletion, caused by the need for society over time to resort to lower-grade, more remote, more difficult to process—and hence more costly—mineral resources. These shortages are different. First, they are not common. Indeed, it is difficult to identify any past shortages due to mineral depletion. In addition, should they occur, they are likely to persist for decades or centuries and so threaten the material foundations of modern civilization. They are unlikely to arise unexpectedly or suddenly, but rather slowly, stalking society over decades as real prices march higher and higher.

There are two common ways of thinking about mineral depletion—the fixed-stock paradigm and the opportunity-cost paradigm. The first focuses on the fact that the supply of any mineral commodity is a finite fixed stock. It allows us to assess future availability by calculating the life expectancies of reserves, resources, or the resource base under likely demand scenarios. Despite its logical appeal, the fixed-stock paradigm suffers from several serious shortcomings. It fails to take account of recycling and secondary production, the substitution of more abundant and possibly renewable resources for less abundant resources, and the likelihood that economic depletion would occur before the physical exhaustion of reserves, resources, or the resource base.

As a result, economists and many others now favor the opportunity-cost paradigm. This approach focuses on economic rather than physical measures of availability, and in particular on the changes over time in what society has to give up (typically measured by real prices) to obtain another ton of steel or of other mineral commodities.

Historical information on commodity prices indicates that mineral depletion has not created serious problems. Abstracting from short-run fluctuations, the real prices for mineral commodities have either remained more or less the same or have trended downward since the latter half of the nineteenth century.

As for the future, we know that merely extrapolating past price trends is unlikely to provide reliable forecasts. A better approach employs the cumulative availability curve, which shows the total quantities of a given mineral commodity available (in the sense that they are profitable to produce) at various prices over all time. It considers separately the influence of three groups of factors on the future supply and demand for mineral commodities and hence their prices. The first group determines the shape of the curve; the second, the pace society moves up the curve; and the third, shifts of the curve over time.

The future hinges largely on the outcome of a race between the cost-increasing effects of depletion (governed by the shape of the curve and how quickly society moves up it) and the cost-reducing effects of new technology (governed by the extent to which the curve shifts downward). The speed at which society will move up the curve and the extent to which the curve will shift downward in the future are unknown and presumably unknowable. This has led some, including one of us (Tilton, 2003), to conclude that assessing the future threat of depletion and predicting the future availability of mineral commodities are simply not feasible. While many claim otherwise—insisting that abundance will continue indefinitely or that serious shortages are inevitable—these prophets are asking the rest of us to buy into their faith that new technology either will or will not offset the future adverse effects of depletion.

Over the past decade, however, we have come to realize that in certain instances knowledge of the shape of the cumulative availability curve by itself can provide powerful insights regarding future availability. For copper and most mineral commodities, the shape of the curve is unknown, but not unknowable. The shape of the curve depends on geological events that occurred in the past, and not on events that may or may not take place in the future. If society were willing to make the investment, it could learn a great deal about the nature and incidents of sub-economic resources.

For a few mineral commodities, we now have estimates for the shape of their cumulative availability curves. Where these curves rise gradually and eventually become quite flat, we can estimate the maximum long-run price under any plausible future demand scenario. For lithium, for instance, almost unlimited supplies are available from seawater at a maximum price of no more than 7–10 dollars per pound of lithium carbonate.

Not all mineral commodities, of course, may have such benevolent cumulative availability curves. Those whose curves rise steeply or have discrete breaks are more exposed to the threat of depletion. There is, however, an important asymmetry between mineral commodities with gradually rising cumulative availability curves and those with steep slopes or with discrete breaks. With the former, if the relatively flat portion of the curve occurs at prices close to current prices, we can confidently conclude that depletion will not be a serious threat. With the latter, on the other hand, we cannot confidently conclude that depletion will be a problem. This is because we have no way of knowing to what extent new innovations and technology will offset the upward pressure on costs caused by depletion.

Several other interesting, and in some cases counter-intuitive, implications flow from the opportunity-cost paradigm and the cumulative availability curve as well. First, contrary to popular opinion, population growth may increase rather than decrease the future availability of mineral commodities. It is true that population growth increases demand and accelerates the pace at which society moves up the cumulative availability curve. However, it also adds to the stock of human capital creating the new technologies that shift the curve downward. So population growth may actually lead to lower commodity prices.

Second, until the recent surge in Chinese demand, the United States and other developed countries with some 20 percent of the world's population consumed some 80 percent of its non-renewable mineral resources. To many this seems unfair. However, the reasoning just used to suggest that population growth may enhance rather than inhibit future availability can be employed here as well. Consumption in developed countries does accentuate the move up the cumulative availability curve, but the wealth it creates also supports the innovative activities that shift the curve downward. The vast majority of the new technologies promoting mineral availability over the past century or two have come from the United States and other developed countries. This means that developing countries may very well have benefited from—rather than suffered from—the tremendous mineral consumption in the developed world over the past two centuries. We know that today developing countries have access to mineral commodities that are no more expensive and often cheaper than they were when the United States and other industrialized countries started to develop.

Third, the cumulative availability curve highlights the fact that the world is in transition and evolving, and simply cannot be preserved in its present state. Those who claim that we cannot rely forever on conventional oil and gas and on rich mineral deposits—copper deposits blessed with grades of 0.8 percent and higher— are surely right. But this, by itself, tells us little about the future availability of mineral commodities and has little relevance for the future welfare of the human race. The relevant question is whether society can satisfy the needs currently served by these high-quality mineral assets from other resources at real prices close to or below current prices.

Finally, the cumulative availability curve makes it abundantly clear that the major weapon in society's arsenal in its eternal struggle with mineral depletion is human ingenuity, specifically the ability to innovate and develop new technologies. We are not passive spectators watching helplessly as external forces over which we have no control determine the outcome of the race between the cost-increasing effects of depletion and the cost-reducing effects of new technology. Should depletion start to drive real mineral prices upward, society can respond. In particular, it can devote more of its talents and resources to finding new and better ways of meeting its mineral needs.

Other suggested policies, such as promoting conservation, recycling, and the substitution of renewable for non-renewable resources, slow the rate at which society moves up the cumulative availability curve, but they do not shift the curve downward. Shifts in the curve require innovation and new technologies. Slowing the movement up the curve can postpone the day of reckoning. Shifting the curve downward can eliminate the day of reckoning indefinitely and produce a future where mineral commodities are more, rather than less, available.[12]

Notes

1 This chapter draws heavily from Tilton (2003), Tilton (2006), and Yaksic and Tilton (2009).
2 The assumptions are: (1) the mine maximizes the net present value of its current and future profits; (2) the mine is a competitive producer and so has no control over the price it receives; (3) there is no uncertainty, so the mine knows the size and nature of its resource stock as well as current and future costs and prices; (4) the mine's output

is not limited by existing capacity or other constraints, so it can produce as little as nothing or as much as its entire remaining resource stock during any time period; (5) the mine's resource stock is uniform so that grade and other qualities do not vary; (6) technology does not change.

3   The term "resources" has two different meanings—a measure of current plus potentially feasible reserves, the definition here; and the mineralized material from which mineral commodities are extracted, the definition noted at the beginning of this chapter. Which of these meanings applies when the term is used should be clear from the text.

4   The earth's crust is the outermost layer of the earth. It varies in thickness from 8 to 70 kilometers. It is thinnest beneath the oceans and does not include the oceans or other bodies of water.

5   Vincent E. McKelvey, a geologist and former director of the US Geological Survey, developed a widely used resource classification scheme around what has become known as the McKelvey box (McKelvey, 1973). For more on the current classification system of the US Geological Survey, based on McKelvey's early work, see Appendix C of *Mineral Commodity Summaries 2015* (US Geological Survey, 2015a) or earlier editions of this annual report.

6   Marginal producers are those whose extraction costs plus user costs just equal the market price. In Figure 5.2, for example, the marginal producer is Mine G when the market price is P_m'. Marginal reserves are those of the marginal producer. Under certain conditions, user costs or Hotelling rents reflect the value of marginal reserves in the ground. It is also worth noting that the three measures of opportunity costs can move in different directions. The extraction costs of marginal producers can be rising while the value of their reserves is falling, or vice versa. And, if extraction costs are rising while the value of reserves is falling, the real price (the sum of the two) can be rising or falling. This is because the three measures reflect different aspects or sources of scarcity. User costs focus on the availability of the resource in the ground. Extraction costs take into account the production process and its impact on availability. Prices reflect the combined effect of both and hence the availability of the mineral commodity after extraction and processing.

7   For a more complete discussion of the shortcomings of the economic measures of resource scarcity, see Tilton (2003, ch. 3).

8   For surveys of this literature, see Tilton (2003, ch. 4) and Krautkraemer (1998).

9   Tilton and Skinner (1987) first suggested the cumulative availability curve. Originally, it was called the cumulative supply curve, but is now commonly referred to as the cumulative availability curve in part to avoid confusion with the traditional supply curve.

10   According to the United Nations (2013), world population growth, after changing little over most of human history, started increasing in the seventeenth or eighteenth century. It reached a peak of 2.07 percent per year around 1970. By 2010 it had declined to 1.20 percent. The United Nations expects the decline to continue, reaching 0.51 percent by the middle of this century and 0.11 percent by its end.

11   Demographers who take into account the influence of rising educational levels on fertility rates argue the United Nations projections are too high. They believe world population is likely to peak at 9.4 billion around 2070 and then decline to around nine billion by the end of this century (Lutz *et al.*, 2014).

12   Efforts to promote conservation, recycling, and the substitution of renewable for non-renewable resources if pushed too far, it is worth noting, may actually be counterproductive in terms of promoting availability if they divert resources away from the innovative activities that shift the cumulative availability curve downward.

References

Aguilera, R.F., Eggert, R.G., Lagos, G. and Tilton, J.E., 2009. Depletion and the future availability of petroleum resources. *Energy Journal*, 30, 141–174.

Ayres, R.U., 1993. Cowboys, cornucopians and long-run sustainability. *Ecological Economics*, 8, 189–207.

Barnett, H.J. and Morse, C., 1963. *Scarcity and Growth*, Johns Hopkins for Resources for the Future, Baltimore, MD.

Beckerman, W., 2003. *A Poverty of Reason: Sustainable Development and Economic Growth*, The Independent Institute, Oakland, CA.

BP, 2015. *BP Statistical Review of World Energy.* Available at www.bp.com/statisticalreview

Campbell, C.J., 1997. *The Coming Oil Crisis*, Multi-Science Publishing Company, Brentwood, UK.

Deffeyes, K.S., 2001. *Hubbert's Peak: The Impending World Oil Shortage*, Princeton University Press, Princeton, NJ.

Deffeyes, K.S., 2005. *Beyond Oil: The View from Hubbert's Peak*, Farrar, Straus and Giroux, New York, NY.

Erickson, R.L., 1973. Crustal abundance of elements, and mineral reserves and resources, in Brobst, D.A. and Pratt, W.P., eds., *United States Mineral Resources*, Geological Survey Professional Paper 820, Government Printing Office, Washington, DC, 21–25.

Gordon, R.B., Koopmans, T.C., Nordhaus, W.D. and Skinner, B.J., 1987. *Toward a New Iron Age? Quantitative Modeling of Resource Exhaustion*, Harvard University Press, Cambridge, MA.

Hotelling, H., 1931. The economics of exhaustible resources. *Journal of Political Economy*, 39 (2), 137–175.

Hubbert, M.K., 1962. *Energy Resources: A Report to the Committee on Natural Resources*, National Academy Press, Washington, DC.

Hubbert, M.K., 1969. Energy resources, in Cloud, P., ed., *Resources and Man*, W.H. Freeman, San Francisco, CA, 157–239.

Jevons, W.S., 1865. *The Coal Question*, Macmillan, London.

Jordon, B.W., Eggert, R.G., Dixon, B.W., and Carlsen, B.W., 2015. Thorium: crustal abundance, joint production, and economic availability. *Resources Policy*, 44, 81–93.

Kesler, S.E., 1994. *Mineral Resources, Economics and the Environment*, Macmillan, New York, NY.

Krautkraemer, J.A., 1998. Nonrenewable resource scarcity. *Journal of Economic Literature*, 36, 2065–2107.

Lee, T. and Yao, C.-L., 1970. Abundance of chemical elements in the earth's crust and its major tectonic units. *International Geological Review*, 12 (7), 778–786.

Lomborg, B., 2001. *The Skeptical Environmentalist*, Cambridge University Press, Cambridge.

Lutz, W., Butz, W.P. and Samir, K.C., 2014. *World Population & Human Capital in the Twenty-First Century*, Oxford University Press, Oxford.

McKelvey, V.E., 1973. Mineral resource estimates and public policy, in Brobst, D.A. and Pratt, W.P., eds., *United States Mineral Resources*, Geological Survey Professional Paper 820, Government Printing Office, Washington, DC, 9–19. This article also appears in *American Scientist*, 60, 32–40.

Meadows, D.H., Meadows, D.L., Randers, J., and Behrens, W.W., 1972. *The Limits to Growth*, Universe Books, New York, NY.

Neumayer, E., 2000. Scarce or abundant? The economics of natural resource availability *Journal of Economic Surveys*, 14 (3), 307–335.

Pinchot, G., 1910. *The Fight for Conservation*, Doubleday, Page and Company, New York, NY.

President's Materials Policy Commission, 1952. *Resources for Freedom, Volume I: Foundations for Growth and Security*, US Government Printing Office, Washington, DC.

Simon, J.L., 1981. *The Ultimate Resource*, Princeton University Press, Princeton, NJ.

Skinner, B.J., 1976. A second iron age ahead? *American Scientist*, 64, 158–169.

Tilton, J.E., 2003. *On Borrowed Time? Assessing the Threat of Mineral Depletion*, Resources for the Future, Washington, DC.

Tilton, J.E., 2006. Depletion and the long-run availability of mineral commodities, in Doggett, M.E. and Parry, J.R., eds., *Wealth Creation in the Minerals Industry: Integrating Science, Business and Education*, Special Publication 12, Society of Economic Geologists, Littleton, CO, 61–70.

Tilton, J.E. and Lagos, G., 2007. Assessing the long-run availability of copper. *Resources Policy*, 32, 19–23.

Tilton, J.E. and Skinner, B.J., 1987. The meaning of resources, in McLaren, D.J., Skinner, B.J., eds., *Resources and World Development*, John Wiley & Sons, New York, NY, 13–27.

United Nations, 2013. *World Population Prospects: The 2012 Revision, Volume 1: Comprehensive Tables*, United Nations, Economic and Social Affairs, New York, NY. Available at: http://esa.un.org/unpd/wpp/Documentation/pdf/WPP2012_Volume-I_Comprehensive-Tables.pdf.

US Geological Survey, 2015a. *Mineral Commodity Summaries 2015*, US Geological Survey, Washington, DC. Available at: http://minerals.usgs.gov/minerals/pubs/mcs/2015/mcs2015.pdf.

US Geological Survey, 2015b. *Historical Statistics for Mineral and Material Commodities in the United States*, US Geological Survey, Washington, DC. Available at: http://minerals.usgs.gov/minerals/pubs/historical-statistics.

Yaksic, A., 2008. Análisis de la disponibilidad de litio en el largo plazo. Unpublished MS thesis, Pontificia Universidad Católica de Chile, School of Engineering, Santiago, Chile.

Yaksic, A. and Tilton, J.E., 2009. Using the cumulative availability curve to assess the threat of mineral depletion: the case of lithium. *Resources Policy*, 34, 185–194.

10 The Environment and Sustainable Development

Over the past half-century, concern over two issues—first, pollution and the environment, and second, sustainable development and intergenerational equity, including global climate change—have emerged to dominate the public discourse regarding growing threats to humanity. The mining and mineral sector, for better or worse, finds itself center-stage in these discussions. This is in part because the extraction of mineral commodities entails the generation of considerable wastes. These wastes damage the land, pollute the air, and contaminate both surface and subsurface water resources. In addition, mining exploits non-renewable resources. For many this raises the question of how modern civilization, with its growing dependence on minerals, will survive once its high-quality resources are gone.

This chapter explores these concerns and public policies for addressing them. The next section focuses on pollution and the environment, the following on sustainable development.

Pollution and the Environment

Mining is well known for its pollution. A century ago this was seldom considered a serious issue. However, as population has grown and pollution from other sources has increased, preserving the quality of our land, water, and other environmental assets has become much more important.

The Concern

We begin by looking at the question: Why is pollution a concern? The answer at first may seem self-evident, but actually it is not so simple. A related question is: How much pollution is optimal from the point of society? One's first response might be: the less, the better. But this implies that zero pollution is the optimal, which upon reflection is clearly not true. Imagine the human condition if absolutely no pollution were allowed.

Land, water, air, and other environmental assets are resources, just like labor, capital, energy, and materials. If the objective function of society is to maximize human welfare (as Chapter 5 proposes), then we should use environmental resources, permitting pollution up to the point where the social benefits flowing from the next unit used just equals the social costs. This does not mean, however, that the marketplace by itself will ensure the optimal use—no more, no less—of environmental resources.

To understand why, we need to define several terms. *Social costs* cover all the costs affecting society as a whole arising from a particular activity. They include the costs that producing firms, and in turn consumers, pay, such as the costs of labor, capital, and materials. The latter are what we call *internalized costs*, precisely because the firm or party that uses these resources pays for them.

Social costs may also include *external costs* or what are often referred to simply as *externalities*.[1] These are costs imposed on members of society other than the firms and consumers that cause them. For example, when a mine pollutes a stream that downstream communities use for drinking water, the increased expense these communities incur in purifying their water reflects an external cost of the mine's operations. When the mine purifies the water it has polluted, due to government regulations or for other reasons, making it unnecessary for downstream communities to do so, what were external costs become internalized. They are now paid for by the mine and are reflected in the price it charges its customers.

External costs, it is worth noting, arise for reasons other than pollution. When firms, for example, bribe government officials and engage in other forms of corruption, much of the costs fall on other members of society.

External costs are a type of market failure, as Chapter 5 notes. They create a number of problems. First, as we saw in Chapter 9, they complicate the use of real commodity prices in assessing the effects of mineral depletion on resource availability. We return to this issue later in this section.

Second, they raise concerns about equity or fairness. Like beauty, equity entails a personal judgment in the sense that reasonable people can disagree over what is and what is not equitable. As a result, economists often try to avoid discussing equity. However, where there are external costs, society is subsidizing the goods being produced. In competitive industries and most others as well this means consumers pay prices that are less than the full social costs of production. Why should society give consumers of these products discounts but not the consumers of other products? Another equity issue concerns who should pay for pollution. Here, again, reasonable people can have different opinions. Still, most subscribe to the view that those who make a mess should clean it up. We learn this lesson as children when our parents make us put away our toys and clean up our rooms.

Third, and most important to economists, external costs give rise to three types of inefficiencies.[2] The first is *allocative efficiency*. When firms do not incur the full social costs of their activities, as just pointed out, their production is subsidized and their prices as a result are too low. This encourages the consumption of these goods over others whose costs are completely internalized.

Figure 10.1 illustrates this situation. It shows the market demand curve (DD) and the market supply curve (SS = PMC) for a good sold on a competitive market. As Chapter 3 points out, the supply curves for individual competitive firms are their marginal cost curves above the minimum point on their average variable cost curve. Since profit-maximizing firms increase output until the costs they incur in

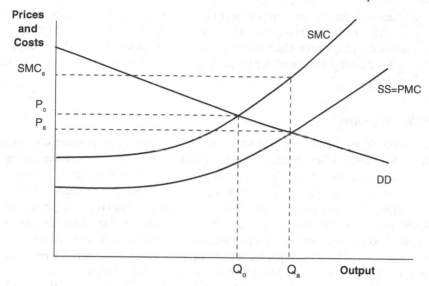

Figure 10.1 Actual and optimal market conditions with external costs.

producing the next unit just equal the market price, their supply curves reflect their private or internalized costs. So the market supply curve, determined by the horizontal additional of each firm's supply curve, reflects the industry's private marginal cost curve (PMC) rather than its social marginal cost curve (SMC). The latter lies above the former by exactly the external costs at various output levels.

Under these conditions, the actual market price and output are P_a and Q_a. From the perspective of producing firms this is ideal since the cost to them of the last unit produced just equals the price received for that unit. However, from the perspective of society, the benefit to society of the last unit produced equals the market price,[3] while the costs to society (SMC) are much higher. For society the optimal price (P_o) is higher than the actual price (P_a) and the optimal output (Q_o) is lower than the actual output (Q_a). Society can enhance its welfare by allocating fewer resources to the production of these goods and more to other goods and services.

The second is *production efficiency*. It is concerned with the optimal mix of inputs in the production process. When clean water or clean air are free to producers, or just underpriced in terms of their true social value, firms tend to overuse these resources, substituting them where they can for capital, labor, and other inputs for which they must pay their full social value. So, while allocative efficiency focuses on how much we produce of various goods and services, production efficiency focuses on how we produce goods and services.

The third is *dynamic efficiency*. It reflects the optimal creation and diffusion of new technologies. When producers have free access to environmental resources, they have no incentive to develop new technologies that reduce or eliminate their need for these resources. When environmental resources are not free but underpriced, the incentives are insufficient. Over the long run, dynamic inefficiency is without question the most important of the three types of inefficiencies. As the previous chapter highlights, human ingenuity and the creation of new innovations and new technologies constitute the major weapon in society's arsenal for keeping the cost-increasing effects of depletion in check. External costs compromise its effectiveness.

Policy Response

Conceptually the remedy for external costs and the various concerns just discussed is simple: public policy should, through regulations or other means, internalize all external costs. This shifts the industry's private marginal cost curve up until it coincides with the industry's social marginal cost curve. So, in Figure 10.1 the new industry supply curve is now SMC. At the new equilibrium output Q_o, the social costs of the last unit produced, including pollution and any other previously external costs, just equal the social benefits. The pollution associated with this new equilibrium level of output reflects the optimal use of environmental resources and the optimal level of pollution from the point of view of society.

In practice, however, there are challenges to internalizing external costs. In particular, this requires three necessary conditions. First, society must be able to value its environmental resources and to measure external costs. Without a reliable indication of the external costs associated with various activities, policies designed to internalize them are not feasible. Second, society must have the means or policy tools to force firms to internalize their external costs. Third, once governments have the ability to measure external costs and the means to internalize them, they must possess the will to do so.

Valuing Environmental Resources and Measuring Externalities

Of the three conditions necessary to internalize external costs, measuring these costs is perhaps the most problematic. For labor, capital, and many other inputs used in the mining and processing of mineral commodities, market prices provide a reasonable estimate of their social value. However, some inputs, including clean water, pristine wilderness, indigenous cultures, and biodiversity are not traded on markets. As a result, we have to use alternative means to assess their value.

One approach is to create markets for environmental resources. The US government, for example, in an effort to minimize the cost of reducing acid rain, has for some two decades allowed domestic coal-powered electric utilities to trade sulfur dioxide permits. These permits are bought and sold on an exchange. Companies that can reduce their sulfur dioxide emissions at a cost below the going permit price have an incentive to sell their permits. The opposite is true for companies with abatement costs above the price of permits. As a result, the market price for permits reflects the lowest possible cost to society of reducing sulfur dioxide emissions by one additional ton. More recently, a number of countries have adopted similar schemes for trading carbon dioxide permits to meet their Kyoto commitments and to curb greenhouse gas emissions. If, and only if, the total number of permits allowed reflects the socially optimal emissions of sulfur dioxide or carbon dioxide, the permit price provides a good measure of the social value of these environmental resources.[4]

Another approach relies on inferences from consumer behavior to value environmental assets. For example, differences in the market value of similar houses close to and away from an airport provide estimates of what people are willing to pay to avoid noise pollution. Similarly, the expenses people incur in traveling to and from lakes, streams, and other recreational facilities provide a minimum price for the value they place on such natural resources.

A third approach is *contingent valuation*. This technique arose primarily to value environmental goods prized in part or all for their non-use or existence value. Many people, for example, place a positive value on the wilderness in Alaska or the indigenous cultures of Amazonian tribes in Brazil, even though they never plan to visit Alaska or Brazil. Such non-use value is in practice particularly difficult to measure with any accuracy. Contingent valuation attempts to do so by asking people, typically in a structured manner following established guidelines, how much they would pay to preserve these resources. This technique has been used with some frequency in the United States and elsewhere for resource damage assessments in both legal cases and policy inquiries.

Contingent valuation is, however, quite controversial for two quite different reasons. The first is the methodology, which is not based on actual behavior. Those surveyed do not actually have to pay what they say they would pay. Experts agree that poorly designed contingent valuation studies—for instance, ones where the respondents are inadequately informed or the questions are leading—can produce highly misleading results. There is much less agreement, however, about the reliability of well-designed studies. Here, the concern is that contingent valuation does not produce the information it claims to—namely, the true willingness of people to pay.

The second line of criticism questions the use of contingent valuation for more fundamental philosophical or ethical reasons. Here, it is argued, even if this approach does elicit reliable responses, adding the willingness to pay across all members of society is not the appropriate way to value social goods with large non-use values. Biodiversity, pristine wilderness, native cultures, and other such human resources are public goods like national defense and public health. While society could use contingent valuation to estimate the social value of national defense, this is not how governments make budgetary decisions regarding the amount to spend on defense or other public goods. Rather, such decisions are made through a political process that for better or worse reconciles the various competing interests of society, and not an artificially constructed simulation of the marketplace. These two processes are quite different and can produce very different results as well.

The political process normally allows for public debate. This provides individuals and organizations with the opportunity to make their own views known, and to become better informed about the issues and more aware of the views of others. The political process also accommodates the fact that members of society may make decisions about the purchase of private goods in a different way using different criteria than decisions about public goods. As citizens, for example, individuals may support policies (such as higher taxes for education) for the good of society as a whole that have no impact or even a negative impact on their own welfare. The political process, it is argued, is much more likely to take into account and reflect such support than contingent valuation.

To be sure, both approaches give the wealthy a greater say in the provision of public goods (since the rich have more dollars to spend on questions posed in contingent valuation studies and more dollars to spend influencing the political process). With the political process, however, every citizen no matter how poor ultimately has one vote in electing public officials. This may leave the poor feeling less disenfranchised.

This review of various approaches for valuing environmental resources is far from complete, yet it suffices to show that measuring the full social costs associated with production and use of mineral commodities is complex and difficult. While our tools in this area have advanced considerably in recent years, much progress is still needed before reliable measures of environmental resource values are available, particularly for those resources with substantial non-use value.

Means

Once the external costs are measured, governments need the means or policy instruments to force producers to internalize these costs. For some countries in early stages of development, the required institutional capacity may still be lacking. But for most countries this is not a problem. The means clearly exist. Indeed, the ongoing debate is over which set of policy tools governments should rely upon.

Command and control regulations require firms to meet certain standards or to use particular pollution abatement equipment. An alternative set of policy instruments relies on *economic incentives* to alter firm behavior. For example, governments may impose a tax on producers, often referred to as a Pigovian tax, after Arthur Pigou, the British economist who first recommended this solution to the problem of external costs. Like regulations, such taxes shifts the industry's supply curve upward and causes output to fall toward its socially optimal level.

The government can impose the tax on a firm's output (copper production) or on its use of environmental resources (arsenic emissions, water pollution, and so on). However, taxing the use of environmental resources is far preferable, since it is pollution (the bad) not production (the good) that society wants to discourage. A tax on pollution but not output will encourage firms to find ways to produce more output with less pollution.

The government can also alter economic incentives by privatizing the property rights for environmental assets. We saw this earlier with the creation of tradable permits for sulfur dioxide and carbon dioxide gases. Firms now can buy and sell these assets, which provides an additional economic incentive to use them as efficiently as possible.

Historically, governments have relied primarily on command and control regulations rather than on taxation, tradable permit schemes, and other economic incentives. The past several decades, however, have seen growing support for economic incentives, as they promote greater efficiency in the use of environmental resources, particularly dynamic efficiency.

The Will

Finally, governments must have the will to force firms to internalize external costs. Normally, this is the case. However, there are a few important exceptions. The first arises with state mining companies. Where the government is both the owner and regulator of mining companies, a conflict of interests may arise. The devastating environmental damage that took place under central planning and state ownership in countries such as Russia and Ukraine during the latter half of the twentieth century is a striking example.

Artisanal and small-scale mining (see Box 10.1) is another example. These mines are highly inefficient. They may leave ore in the ground that better-run operations would exploit. They are more dangerous, and some employ children. Per unit of output they are far more damaging to the environment. Many gold operations, for example, discharge mercury into the surface and groundwater, harming their own workers as well as others. Acid mine drainage, soil erosion, deforestation, and river silting are also common problems. Sites are typically abandoned with little or no reclamation. In many respects, artisanal and small-scale mining is resource exploitation at its worst, but it provides a subsistence existence to many millions of individuals with few, if any, alternatives. Governments, as a result, are reluctant to close down these operations or even in many cases to enforce existing environmental and safety regulations.

Box 10.1 Artisanal and Small-Scale Mining

Artisanal and small-scale mining is carried out, often illegally, by individuals, families, and slightly larger groups using the simplest and most primitive equipment. While data on the number of men, women, and children working in this sector are difficult to come by, available estimates suggest the number is huge—somewhere between 13 and 20 million. Individuals whose livelihoods depend indirectly on artisanal and small-scale mining are estimated at about 100 million (ICMM, 2014; Darby and Lempa, 2014). These figures suggest that artisanal and small-scale mining employs more people than larger and more formal mining operations.

The World Bank (Barry, 1996, p. 3) has estimated that artisanal mining accounted for 20 percent of the gold, 40 percent of the diamonds, and nearly all the gemstones produced in Africa at the end of the twentieth century. Somewhat less than half of Brazil's gold production came from such operations, down from 70 percent in earlier years. In addition to gold, diamonds, and gemstones, artisanal miners produce copper, platinum, silver, tin, zinc, and coal.

Still another exception arises when the external costs are global in nature. For an example we need only consider greenhouse gas emissions and climate change. Developing countries claim that they did not create the problem, and should not now have to slow their growth to reduce greenhouse emissions. The developed world argues that selective cutbacks by only a few countries will not be effective. Another complication: some countries may benefit from global warming, or believe they may benefit, and so are not strongly inclined to support efforts to abate climate change. Finally, all countries are reluctant to bear more of the costs than they believe is their fair share, which invariably is less than what other countries suggest. In such situations the government's will to act is often inadequate.

Past Trends and Future Prospects

The pollution arising from the production and use of mineral commodities, as we have seen, is not necessarily a public policy concern. As long as mining firms, and ultimately their consumers, pay the full social costs of pollution, the marketplace without government intervention should ensure that valuable environmental resources are used optimally.

In practice, the social costs of pollution can be completely internalized, completely external, or a combination of the two. Pollution becomes a policy issue when all or a significant portion of its costs are external. This, as we have seen, raises equity issues and gives rise to several different types of inefficiencies. So the policy problem is not pollution *per se*, but rather external costs and the failure of the users of these valuable resources to pay their full social costs.

The past century has witnessed two offsetting tendencies regarding the external costs incurred in mining. First, as world population has grown, the value that society attaches to its environmental resources has clearly increased. In the late 1800s, when mining companies in Colorado and Montana dumped their waste into nearby streams, few were troubled. The number of pristine streams remaining in the western United States seemed endless. Today, of course, we attach a high value to clean surface water everywhere. This means that society now places a much higher value on the environmental resources used in mining and mineral processing.

Second, and in large part the result of this first trend, the United States and most other countries have in recent years increasingly imposed regulations, taxes, and other economic incentives to ensure mining companies internalize more and more of their external costs.

These two offsetting developments raise the intriguing question of whether the external costs associated with producing an ounce of gold or a pound of copper have risen or fallen over the past 50 to 100 years. While there is no way of knowing for certain, the external costs per unit of output likely have fallen for most mineral commodities, and for many they have probably fallen substantially. When we look at the numerous mining Superfund sites that pock the western United States, for example, they are all the legacy of mining many decades ago. Given the growth in the total output of mineral commodities, however, the absolute value of the associated external costs may well have risen.

The general public tends to paint the whole mining industry with the same brush when it comes to its impact on the environment. There are, though, important differences. As we have just noted, mining today seems much less polluting per unit of output than was the case a century ago. And, today the major multinational mining companies pollute far less per unit of output than artisanal and small-scale miners, where the government's will to force the internalization of external costs, as we have just seen, is often lacking.

Turning from the past and the present to the future, the picture seems reasonably promising. The key policy issue is to ensure all costs are internalized. We know that governments have the means to do this. The challenge is deciding when to use command and control regulations and when to use Pigouvian-type taxes and other economic incentives. We also know that governments generally have the will to internalize outstanding external costs despite a few important exceptions. For mining, artisanal and small-scale mining along with state mining companies pose the biggest challenges. For society as a whole, greenhouse gas emissions and climate change are of great concern.

Probably the biggest challenge that remains is finding reliable ways to value environmental assets and their associated external costs, particularly those assets with significant non-use value. Environmental economists and others have made impressive progress in this area over the past several decades, and are likely to continue to do so. What ultimately we need are valuations upon which reasonable people will agree regardless of the strength of their own personal concerns regarding the environment.

External Costs and Measuring Long-Run Availability

Before shifting our attention to sustainable development, there is a digression we must take to address the concern raised in Chapter 9 that external costs undermine the validity of using trends in the real prices of mineral commodities to assess the availability over the long run of mineral commodities. Here, there are two frequently made claims that require investigation. First, in the presence of external costs, mineral commodity scarcity may be growing even though real prices are trending downward. Second, even if long-run trends in real prices remain stationary or fall in the future, as they have in the past, shortages of mineral commodities are likely to arise as society curtails their production in response to growing environmental concerns.

The first of these claims calls into question the conclusion reached in Chapter 9 that mineral commodities over the past century or two have not become less available. Here again the problem is not the pollution associated with mining, but the external costs. Market prices reflect the internalized costs of production. What we would like to have are prices that reflect full social costs.

Where there are external costs, reported prices are too low. Assume, for example, that the external costs incurred in the production of a given mineral commodity are a constant 50 percent of the internalized costs. Then prices would be too low by one-third. Note, however, that this need not affect the trend. As long as the external costs remain equal to 50 percent of internal costs, the trend over time in reported prices (reflecting just internalized costs) will be the same as the price trend that reflects full social costs.

So what matters is how the share of external costs in total social costs changes over time. This share, as we saw earlier, changes in response to two offsetting forces. The first—the increasing value society places on its environmental resources over time—tends to increase the share of external costs. The second—the growing regulations and other measures that governments impose to force producers to internalize their external costs—has the opposite effect.

Those who argue that trends in real prices are a biased measure of availability and as a result significantly underestimate the extent to which mineral commodities have become scarcer over the past century or two—or have significantly overestimated the extent to which they have become more abundant—are correct only if: (a) external costs account for an appreciable share of the full social costs; and (b) the share of external costs is growing over time. While these conditions may hold in a few cases, there is little evidence indicating that external costs as a share of total social costs per unit of output have increased for most mineral commodities.

The second frequently made claim about the future is that growing concerns about the environment will limit mineral production and create scarcities, regardless of how the costs and prices of mineral commodities evolve in the future. Two lines of reasoning underlie this prediction.

First, for various reasons society fails to require producers to internalize adequately their external costs. Eventually, public frustration arising from this failure forces governments to curtail or completely eliminate mineral production. The likelihood of this scenario, however, seems quite low, given that governments have with very few exceptions shown a clear willingness over time to force the internalization of external costs.

Second, many contend that biodiversity, indigenous cultures, and pristine wilderness are examples of environmental assets and other social goods that are simply incompatible with the extraction of mineral commodities. Where this is true, internalizing the costs of these social goods does more than reduce the optimal output of mineral resources—it reduces it to zero. In effect, mining is prohibited where it threatens such social goods.

Does this not mean growing mineral scarcity? The answer is: not necessarily. For years public policy has prohibited mining in certain areas—cities, national parks, and military reservations, for example. Moreover, the total size of such areas has expanded greatly over the past century, while simultaneously the availability of many mineral commodities has increased. So the protection of social goods incompatible with mining is possible without necessarily causing scarcity, though clearly the more territory withdrawn from mineral extraction the greater the challenge for new technology in the struggle to keep mineral costs and prices from rising.

The issue for public policy, it is important to highlight, is not between choosing biodiversity, pristine wilderness, and indigenous culture on the one hand or the availability of mineral commodities on the other. It is not an either/or issue, or a case of black or white, but rather a question of the appropriate balance. How much biodiversity, wilderness, and indigenous culture does society want to preserve? As the amount increases, so does the price to society in terms of the long-run mineral availability sacrificed. At the same time, as the amount increases, the additional or marginal benefits to society fall, assuming the most valuable sites for biodiversity, wilderness, and indigenous culture are selected for protection first.

This means that public policy should continue to preserve these social goods, and exclude mining as necessary, up to the point where the marginal costs (in terms of the resource availability sacrificed) just equals the marginal benefits to society. Such a policy may or may not create scarcities of mineral commodities, but even if it does, the policy is still optimal, promoting the welfare of society as a whole.

Some economists and policy analysts (Krutilla and Fisher 1975; Dasgupta *et al.* 1999) go even further and urge a cautionary policy, one that requires governments when weighing the benefits and costs to take into account the fact that once mining or other activities destroy such social goods, the damage is often irreversible. Moreover, as population and per capita income increase over time, the demand for these goods is likely to grow more rapidly than the demand for most other goods. Unlike other commodities, it is difficult or impossible to produce goods widely considered as close substitutes for biodiversity, indigenous cultures, and pristine wilderness. Such concerns, coupled with the vast quantities of resources that are close to being economic and that are known to exist for many mineral commodities, suggest that a prudent policy at least for the present would preclude mineral development wherever important social goods are threatened.

This, however, does not mean that we have to abandon using trends in real prices to measure trends in the availability of mineral commodities. This would be the case only if the external costs of mineral production were over time accounting for a larger and larger share of the total social costs. And, there is little to suggest this is so.

Sustainable Development and Intergenerational Equity

Sustainable development and sustainability are terms that mean different things to different people. The World Commission on Environment and Development (1987), better known as the Brundtland Commission after its chair, in its report *Our Common Future* is widely credited with introducing the term "sustainable development" into the public lexicon. It defines sustainable development as "development that meets the needs of the present without compromising the ability of future generations to meet their own needs."

Since then, many other definitions have emerged. For some, sustainable development means protecting a particular ecosystem, for others preserving biodiversity, for still others protecting an indigenous culture or a local community from the development of a nearby mine. Then there are those who see sustainable development as helping a mining community remain economically viable after the ore is gone and the mines are closed. In yet another use, sustainable development is the equitable distribution of income, goods, and resources among different countries and people today, and so is void of any inter-temporal dimension.

Eggert (2013) distinguishes between sustainability and sustainable development. He defines sustainability as a one-dimensional concept with three variants—environmental sustainability (sustaining environmental quality and the stock of natural resources), economic sustainability (improving per capita income and other measures of human welfare), and social and cultural sustainability (promoting social and cultural development, including fairness in the allocation of the benefits and burdens of economic activities).

In contrast, sustainable development for him is multidimensional. It entails the simultaneous pursuit of all three variants of sustainability. Eggert's definition of sustainable development is quite broad and encompasses many of the various definitions that others have advanced. It does, however, lead to a lot of ambiguity. This is because many actions advance sustainable development along one dimension while reducing it along another. Artisanal mining, for example, may increase economic development, but given its negative impact on the environment few would argue it contributes to sustainable development.

So in this chapter we use the terms sustainable development and sustainability—which we treat as synonyms—more narrowly. Specifically, we define sustainable development or more simply sustainability to mean that the present generation behaves in a manner that does not preclude future generations from enjoying a standard of living at least comparable to its own. This definition is fairly common among economists. Like the original definition of the Brundtland Commission, it emphasizes intergenerational equity and has a macro orientation, focusing on the welfare of society as a whole rather than the well-being of a particular ecosystem or local community.

We begin by examining mining and sustainable development from the perspective of the world as a whole. We then explore sustainability from the perspective of individual mineral-producing countries.

The Global Perspective

We are all consumers of mineral commodities. Indeed, modern civilization depends on the consumption of tremendous quantities of these non-renewable resources. For those who adhere to the fixed-stock paradigm, this raises a serious, indeed insurmountable, challenge. For as we saw in Chapter 9, according to this view of depletion, the world eventually must run out of its physical stocks of mineral resources. This means that the world as we know it now, with its dependence on mineral commodities, simply is not sustainable.

Fortunately, as we also saw in Chapter 9, the fixed-stock paradigm suffers from several serious flaws. With the more useful opportunity-cost paradigm, the threat is not physical but economic depletion. Should new technology fail to offset the cost-increasing effects of depletion, current consumption would require future generations to pay more for their mineral commodities, which in turn might force them to accept a lower standard of living.

Upon some reflection, however, the link between mineral production and consumption today and sustainable development seems quite tenuous. This is in part because, again as we saw in Chapter 9, there is not a strong negative correlation between mineral consumption today and future availability. Over the past century, consumption has expanded substantially, while scarcity measured by long-run trends in real prices shows little or no tendency to increase.

Moreover, even if mineral consumption were to produce rising prices and less availability, this would not necessarily condemn future generations to lower living standards. This is because their standard of living will depend on all the assets the current generation passes on. In addition to mineral assets, these include physical assets (houses, factories, schools, office buildings, roads, bridges, and other infrastructure), human capital (a healthy and well-educated population), other natural assets (a clean environment, pristine wilderness, rich biodiversity, agricultural resources), political and social capital (stable and democratic government, a well-developed legal system, a tradition of resolving conflict by peaceful means, and other institutions), cultural assets (literature, music, art, dance), and of course knowledge capital (technology, science).

So sustainable development is quite possible even with the declining availability of mineral commodities. This simply requires that the current generation invest sufficiently in the other assets it leaves to future generations to offset the negative impact of increasing mineral scarcity. Indeed, some scholars suggest that even the complete physical depletion of non-renewable resources may not threaten sustainable development, though this is probably a bit of a stretch (see Box 10.2).

Box 10.2 Weak Versus Strong Sustainability

Going one (big) step further, some economists (Solow 1974; Hartwick 1977; Dasgupta and Heal 1979) have argued in favor of what is called *weak sustainability*. This is the view that sustainable development is possible with the complete exhaustion of non-renewable mineral commodities. It is based on models that assume the substitution of other inputs for non-renewable mineral resources is possible in the production of all critical goods. Critics of this view (Daly 1996; Ruth 1995; Neumayer 2000) contend that this assumption defies the laws of nature. They argue instead for *strong sustainability*, which allows for some substitution but not the complete elimination of mineral commodities in the production of goods and services. Not surprisingly, they find the complete exhaustion of mineral resources incompatible with sustainable development.

However, the debate over strong and weak sustainability, while of some intellectual interest, is of questionable practical relevance. For as pointed out in Chapter 9, physical exhaustion is not the issue. We will not literally run out of resources. Scarcity may push the costs of some mineral commodities sufficiently high to preclude their widespread use, but resources will remain in the ground, and so will be available at some price. Higher prices, as we have seen, may increase the challenge of achieving sustainable development, but do not preclude it.

On the other hand, increasing mineral availability, though it may make sustainable development somewhat easier to achieve, certainly does not ensure it. A generation that fails to invest in new technology, to improve the environment, and to enhance human capital in order to save its stock of mineral resources for future use is not likely to achieve sustainable development, nor to earn the gratitude of future generations.

All of this suggests that the pace of mineral extraction by itself does not have a great impact on future generations and sustainable development. Much more important is how much of the wealth that it creates the current generation devotes to its own consumption and how much it invests in human capital, new technology, and other assets. Perhaps of even greater importance is how much of the wealth that it inherits from earlier generations and how much of the potential wealth it could create, the current generation squanders on wars, corruption, needless mismanagement, and other wealth-reducing activities.

The Country Perspective

For mineral-producing countries, sustainable development and intergenerational equity raise another important issue. Chile, Indonesia, Botswana, Australia, and other mineral-rich countries are today mining valuable domestic reserves and often using the proceeds to sustain a higher standard of living than presumably would otherwise be possible. How do they ensure that subsequent generations can maintain that standard of living after their rich deposits are gone and comparative advantage in mining moves elsewhere? To address this question, we need to explore green accounting and Hartwick's rule.

Before doing so, however, it is important to highlight that this concern is quite different than that of the previous section. From a global perspective, the threat to sustainable development is that depletion will force future generations to pay more for their mineral commodities. This, unless offset by other developments, reduces future living standards. So here the focus is on commodity prices and whether or not they are rising over the long term.

From the perspective of mining countries, the concern is that the present generation is running down a valuable national asset (its rich mineral deposits) to support a standard of living that is unsustainable once this asset is depleted. Here, the focus is on trends in the value of the country's mineral assets in the ground rather than commodity prices.

Green Accounting

One of the great economic inventions of the twentieth century was the creation of modern national income and product accounts. Income accounts, such as the well-known gross domestic product (GDP), measure the total income and output of a nation over a given period of time, such as a year. Asset accounts indicate the assets, liabilities, and net worth of a nation at a particular point in time.

These accounts provide a useful report card on a country's economic performance. Is output growing? How is total output divided between investment and consumption? Is the ratio of investment to output rising or falling? How do trends in this ratio compare with those of other countries? Are the country's total assets growing? Are some regions expanding faster than others? How is total income divided between labor, capital, and other resource owners? Such information is not only of intrinsic interest, it is invaluable for public policy.

National income and product accounts do, however, suffer from various shortcomings. For example, many welfare-creating activities that do not involve market transactions are excluded. Unpaid household work and volunteer activities are examples. Changes in the environment are also largely ignored. So China's rapid GDP growth over the past several decades does not reflect the negative effects of the associated pollution.

The shortcoming of particular relevance for our purposes concerns the treatment of natural resources. While national accounts do take into account the production of mineral commodities and their flows through the economy, they completely ignore changes in the value of a country's mineral assets. This is in contrast to their treatment of physical assets, such as plant and equipment, where new investments and the depreciation of existing facilities are recorded.[5] This anomaly is troubling, since mineral resources, just like capital, are often important inputs into the production of goods and services.

This deficiency means that national accounts can show a mining country enjoying strong economic growth, when in fact that growth is unsustainable and actually impoverishing the country. A full reckoning of the costs and benefits would reflect a country simply living off its natural resource wealth.

Green accounting encompasses the efforts over the past several decades in the United States and abroad to supplement the traditional treatment of natural resources (and the environment) in national income and product accounts. A well-designed green accounting system indicates whether or not a mining country is developing in a sustainable manner and not just living off the exploitation of its mineral assets.

Green accounting offers various procedures for estimating the value of reserves in the ground (Nordhaus and Kokkelenberg 1999, ch. 3). As we saw in Chapter 5, the value of mineral resources in the ground is the sum of user costs (or Hotelling rent) and the Ricardian rent associated with existing reserves (see Figure 5.2). Since there are good reasons to believe that user costs are small or non-existent, a country's mineral wealth largely reflects its Ricardian rents.

According to Nordhaus and Kokkelenberg (1999), US mineral wealth changed little over the several decades prior to the publication of their study. This means that the value of reserve additions plus any revaluation of reserves due to price changes more or less offset the value of reserve depletions over time. This provides little support for the view that the United States is in the midst of an unsustainable mineral resource consumption binge, though several decades is perhaps too short a period of time for assessing this proposition.

Another interesting finding of this work: mineral resources account for only a small share of the total wealth of the United States. It values US mineral reserves at only 3–7 percent of the country's physical capital (Nordhaus and Kokkelenberg

1999, p. 104). Adding in human capital and other assets would reduce these figures further.

Perhaps of more interest is the often negative relationship between the mineral wealth of mining countries and the long-run availability of mineral commodities. At first blush one might think that declining mineral availability should reduce mineral wealth, but this is rarely the case. Again, looking back to Figure 5.2, we can see that an increase in a mineral commodity's price, a sign of declining mineral availability, increases the Ricardian rents associated with existing reserves and hence their value.

Alternatively, consider the impact of the new technology the Germans introduced in the early twentieth century that led to the production of artificial nitrates. While this development reduced the costs of producing nitrates, making them less scarce to consumers, it largely destroyed the economic value of the natural nitrate deposits found in northern Chile, creating severe economic problems for this country.

A more recent example—one more favorable for Chile—involves the discovery and development of high-grade, low-cost copper deposits in that country over the past quarter-century. By keeping the world price of copper below what it otherwise would have been, Chile's new mines have had a negative impact on the value of copper reserves in the United States and elsewhere. While the increased mineral wealth enjoyed by Chile may or may not have offset the reductions elsewhere, the new Chilean mines have kept prices below what they would have been, and in the process enhanced the global availability of copper.

Hartwick's Rule

We now return to the question raised at the beginning of this section: namely, how do mining countries ensure that their future generations can maintain the standard of living they now enjoy once their rich mineral deposits are gone? John Hartwick, an economics professor at Queen's University in Canada, addressed this question in an important article (Hartwick 1977). Assuming some degree of substitution between mineral capital and other forms of capital, he shows that a mining country must invest all the Ricardian rents and any user costs realized in producing mineral commodities in other assets, such as physical capital or human capital, in order to offset any resulting decline in its mineral wealth. From this, we have Hartwick's rule. It states that mining countries, to behave sustainably, must ensure that the aggregate value of all the assets they are passing on to future generations does not decline.[6]

Mining countries, it is important to note, need not be reducing the value of their mineral wealth even though they are depleting valuable reserves. Over time the discovery of new reserves and changes in the Ricardian rents associated with existing reserves (due to either increases in mineral commodity prices or to new cost-saving technologies) may more than offset the loss in value due to mining. This means mining countries can achieve sustainable development either by maintaining the value of their mineral reserves or by investing in other assets an amount equal to the reduction in their mineral assets.

Table 10.1 Adjusted net savings as a percentage of gross national income for selected mineral producing countries, 2011, 2012, and 2013

Country	2011	2012	2013
Australia	6.8	12.0	9.3
Bolivia	6.9	5.5	7.3
Botswana	20.4	33.2	29.0
Brazil	6.8	4.3	3.1
Canada	6.7	13.0	6.0
Chile	5.4	−0.2	4.2
China	36.4	35.0	29.5
Guinea	−27.7	−42.8	−50.4
Indonesia	17.1	24.1	22.1
Mongolia	−5.5	13.8	13.9
Morocco	20.2	14.7	13.8
Peru	5.4	7.7	11.3
Russia	7.2	15.3	10.6
South Africa	1.5	0.4	1.2
United States	0.9	7.3	5.0
Zambia	4.0	1.8	–

Sources: World Bank (2013, 2014, 2015).

Sustainable development thus requires that a country's net adjusted savings—that is the difference between the positive investments it is making in certain assets and the negative investments or disinvestments it is making in others (such as its mineral wealth or its environment)—be positive. The World Bank in recent years has attempted to estimate net adjusted savings—or what is often called genuine savings—for a large number of countries. Table 10.1 shows the World Bank estimates for a selected group of mineral-producing countries for the years 2011, 2012, and 2013.

Given the inherent difficulties in estimating adjusted net savings, one should treat these estimates with some caution. Nevertheless, they are of interest, as they suggest that most major mining countries are not exploiting their mineral resources in a non-sustainable manner. The apparent exception among the countries shown in Table 10.1 is Guinea.

Highlights

Concerns over the environment and sustainable development often call attention to mining, as mining along with the processing and use of mineral commodities contribute significantly to global pollution. In addition, many wonder if sustainable development is possible as long as the world depends so heavily on non-renewable resources.

As for the environment, we have seen that the issue is not that mining pollutes. The optimal amount of pollution—the amount that maximizes social welfare—is not zero, but rather occurs when the social costs and benefits of the last unit of pollution are just equal. The need for government action arises only when some or all of the costs of pollution are not internalized and thus not paid for by the producers responsible and ultimately their customers. When there are external costs, innocent third parties bear some of the costs of producing a good even though they reap none of the benefits. This, it is widely believed, is unfair or inequitable. External costs also create inefficiencies, the most troubling of which is they diminish or eliminate the incentives for producers to develop and use new pollution-reducing technologies.

Conceptually, the solution to this problem is straightforward: The government should ensure that all the social costs that mining incurs are internalized. This requires that society can identify and measure the external costs, and as well that governments have the means and will to internalize the costs once identified. While there are a few situations where governments may lack the will, the biggest challenge lies in identifying and measuring the full social costs of mining. Here we have made progress over the past several decades, but still have a way to go.

Over the past half-century the pollution generated per unit of output has declined and in many cases declined greatly for most mineral commodities. Government regulations and other public policies have forced mining companies to internalize an increasing share of their environmental costs. This, in turn, has created the incentives for firms to create new pollution-reducing new technologies. Looking to the future, there are good reasons to believe both of these favorable developments will continue. At the same time, though, the value society attaches to clean air, pristine land, and its other environmental assets will almost certainly continue to rise, causing the optimal level of pollution created by our use of mineral commodities to decline.

External costs have also caused environmentalists and others to question the use of real prices as a measure of resource scarcity. Clearly, where external costs are significant, real prices underestimate the full costs to society of producing and using mineral commodities. However, the direction and significance of the bias that this introduces in trends over time is far less clear. Real prices underestimate growing scarcity (or overestimate increasing availability) only if external costs are accounting for an ever larger share of total social costs. While the rising value society attaches to its environmental resources tends to push the share of external costs up over time, this upward pressure is offset by public policies forcing firms to internalize more and more of their external costs.

Turning to mining and sustainable development, one needs to be aware that there exists many different meanings for sustainable development and sustainability. In this chapter, both terms are used to mean that today's generation behaves in a manner that allows future generations to enjoy a standard of living comparable to its own. With mining, sustainable development raises two particular concerns.

The first is the global perspective that today's generation will impoverish the future by consuming most or all the available high-quality, low-cost mineral reserves. While this is possible, it seems unlikely. The welfare of future generations depends not just on the mineral assets but all the assets the current generation passes on. So any decline in mineral assets can be offset by increases in knowledge and technology assets, physical assets, and human capital. In addition, as often noted, the cost-decreasing effects of new technology can, and in the past have, offset the tendency for depletion to reduce the availability of mineral commodities.

The second concern is the perspective of mineral-rich countries. Here, the focus is not on trends in real prices and mineral commodity availability, but rather on changes over time in the value of a country's mineral assets. Green accounting attempts to identify explicitly these changes and to determine the extent to which a country's growth and per capita income simply reflect the consumption of a dwindling stock of valuable mineral assets. To avoid this type of non-sustainable behavior, countries must follow Hartwick's rule, which requires that mining countries reinvest in other assets an amount equivalent to the Ricardian rents and any user costs realized in the extraction of its valuable mineral resources.

Notes

1   For completeness it is worth noting that externalities may be external benefits rather than external costs. External benefits arise when the activities of individuals or organizations produce benefits to society that they do not capture. For example, when you get a flu shot, this produces benefits for you (the reduced risk of getting the flu) and for others (since your reduced risk of getting the flu means you are less likely to expose others to the flu). Two activities widely presumed to generate considerable external benefits are education and research, as those undertaking and paying for these efforts often are not able to capture all the resulting benefits.
2   See Chapter 4 for a discussion of allocative, production, and dynamic efficiencies.
3   The reason for this is that the market price reflects the reservation price of the marginal consumer (that is, the maximum price that the last consumer is willing to pay) and hence the benefit to that individual and in turn to society.
4   The carbon emissions trading system that Europe has recently adopted provides an unfortunate example of a largely unsuccessful effort, the result of setting the number of allowable permits at too high a level.
5   It is worth noting that a country's net domestic product (NDP) equals its gross domestic product (GDP) minus the depreciation of its physical capital stock. So while GDP figures do not reflect the depreciation of existing facilities, this information is readily available from other national income accounts.
6   Hartwick's rule can be generalized to apply to all types of assets, not just mineral assets. For example, a country that is exploiting its environmental assets must invest in other assets to offset the reduction in value of its environmental assets,in order to behave sustainably. Here, however, the assumption of substitutability between asset types may at some point become questionable. For example, air pollution or water pollution may become so bad that no amount of additional physical capital can adequately compensate future generations.

References

Barry, M., 1996. *Regularizing Informal Mining: A Summary of the Proceedings of the International Roundtable on Artisanal Mining*, Occasional Paper No. 6, World Bank, Industry and Energy Department, Washington, DC.

Daly, H., 1996. *Beyond Growth: The Economics of Sustainable Development*, Beacon Press, Boston, MA.

Darby, S. and Lempa, K., 2014. Advancing the EITI in the mining sector: implementation issues. Unpublished article produced by the World Bank's Oil, Gas and Mining Policy

248   *Public Policy*

and Operations Unit, Washington, DC. Available at http://siteresources.worldbank.org/
EXTEXTINDTRAINI/Resources/advancing_eiti_mining.pdf?resourceurlname
=advancing_eiti_mining.pdf.

Dasgupta, P. and Heal, G., 1979. *Economic Theory and Exhaustible Resources*, Cambridge
University Press, Cambridge.

Dasgupta, P., Mäler, K.G., and Barrett, S., 1999. Intergenerational equity, social discount
rates, and global warming, in Portney, P. and Weyant, J.P., eds., *Discounting and
Intergenerational Equity*, Resources for the Future, Washington, DC.

Eggert, R., 2013. Mining, sustainability and sustainable development, in Maxwell, P., ed.,
Mineral Economics, second edition, Australasian Institute of Mining & Metallurgy,
Carlton, Victoria, 215–227.

Hartwick, J.M., 1977. Intergenerational equity and the investing of rents from exhaustible
resources. *American Economic Review*, 67, 972–974.

ICMM (International Council on Mining and Metals), 2014. Artisanal and small-scale
mining. Unpublished note, London, UK. Available at www.icmm.com/page/84136/
our-work/projects/articles/artisanal-and-small-scale-mining.

Krutilla, J.V. and Fisher, A.C., 1975. *Economics of Natural Environment,* Johns Hopkins
University Press, Baltimore, MD.

Neumayer, E., 2000. Scarce or abundant? The economics of natural resource availability.
Journal of Economic Surveys, 14 (3), 307–335.

Nordhaus, W.D. and Kokkelenberg, E.C., eds., 1999. *Nature's Numbers: Expanding the
National Economic Accounts to Include the Environment*, National Academy Press for
the National Research Council, Washington, DC.

Ruth, M., 1995. Thermodynamic implications for natural resource extraction and technical
change in U.S. copper mining. *Environmental and Resource Economics*, 6 (2),
187–206.

Solow, R.M., 1974. Intergenerational equity and exhaustible resources. *Review of Economic
Studies, Symposium on the Economics of Exhaustible Resources*, 41 (5), 29–45.

World Bank, 2013. *2013 World Development Indicators*, World Bank, Washington, DC.
Available at https://openknowledge.worldbank.org/discover?query=world+developme
nt+indicators&scope=%2F&submit=Go.

World Bank, 2014. *2014 World Development Indicators,* World Bank, Washington, DC.
Available at https://openknowledge.worldbank.org/discover?query=world+developme
nt+indicators&scope=%2F&submit=Go.

World Bank, 2015. *2015 World Development Indicators*, World Bank, Washington, DC.
Available at https://openknowledge.worldbank.org/discover?query=world+developme
nt+indicators&scope=%2F&submit=Go.

World Commission on Environment and Development, 1987. *Our Common Future*,
Oxford University Press, Oxford.

Index

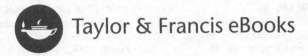

Printed in the United States
by Baker & Taylor Publisher Services